Design Science and Innovation

Series Editor

Amaresh Chakrabarti, Centre for Product Design and Manufacturing, Indian Institute of Science, Bengaluru, India

The book series is intended to provide a platform for disseminating knowledge in all areas of design science and innovation, and is intended for all stakeholders in design and innovation, e.g. educators, researchers, practitioners, policy makers and students of design and innovation. With leading international experts as members of its editorial board, the series aims to disseminate knowledge that combines academic rigour and practical relevance in this area of crucial importance to the society.

Suraj Bhat • Subir Kumar Saha • Vinay Gupta

Technologies for the Base of the Economic Pyramid

Selection and Analysis

Springer

Suraj Bhat
Department of Mechatronics
Mangalore Institute of Technology and Engineering
Moodabidri, Karnataka, India

Subir Kumar Saha
Department of Mechanical Engineering
Indian Institute of Technology Delhi
Delhi, Delhi, India

Vinay Gupta
Department of Mechanical Engineering
IEC College of Engineering and Technology
Greater Noida, Uttar Pradesh, India

ISSN 2509-5986 ISSN 2509-5994 (electronic)
Design Science and Innovation
ISBN 978-981-95-2720-5 ISBN 978-981-95-2721-2 (eBook)
https://doi.org/10.1007/978-981-95-2721-2

This Springer imprint is published by the registered company Springer Nature Singapore Pte Ltd.
The registered company address is: 152 Beach Road, #21-01/04 Gateway East, Singapore 189721, Singapore

To designers and engineers building a better and inclusive world for the other 90%.

Preface

This book is an outcome of the research carried out at the Rural Technology Action Group (RuTAG) at the Indian Institute of Technology (IIT) Delhi. The work originally took shape as a doctoral thesis that explored the challenges and opportunities in designing, selecting, and analysing technologies for the Base of the Pyramid (BoP). Over time, the authors recognised that the insights generated through this research should not remain confined to academic circles. One of the primary motivations for converting the thesis into a monograph has been to make the work more accessible to a larger audience, particularly to practitioners, NGOs, policymakers, and other stakeholders who are directly engaged in technology development and deployment in resource-constrained environments.

Through RuTAG, the authors had the opportunity to witness firsthand the potential of technology to positively impact communities, particularly those from economically weaker sections. At the same time, it was also observed that the process of technology adoption was rarely straightforward. There were several barriers ranging from contextual mismatches to affordability. These experiences underscored the need for a more structured approach to designing, selecting, and analysing technologies suited to the Base of the Pyramid (BoP). This book reflects that effort.

The idea of the BoP is a useful way of understanding this challenge of designing appropriate technological solutions. The global economic distribution is often visualised as a pyramid where a large majority of the world's adult population occupies the wide base. More than half of this population lives with limited income and resources. Their needs are distinct from those at the top of the economic pyramid. In such contexts, technologies must not only be affordable but also contextually appropriate, durable, and user-friendly.

Technology plays an important role in poverty alleviation and the overall development of the society. However, most of the technology is designed for the top of the pyramid. As a result, poverty alleviation efforts through technology interventions often fail or have minimal impact. This is evident when one looks at the number of devices sent as aid to developing countries that are lying unused.

With multiple technologies being developed as a result of developmental efforts worldwide, it becomes important that the right and appropriate technology reaches the user. The critical question is no longer just *what* to invent, but *which* technological solutions are more suitable for a given context. This book attempts to address that gap. Drawing from design literature and field experiences, this book identifies key technology attributes relevant to BoP contexts. Structured approaches to compare and select technologies are proposed and illustrated through case studies. While the examples focus primarily on mechanical systems, the broader principles of evaluation and contextual analysis are widely applicable.

In summary, this book provides a framework for the selection, analysis, and optimisation of technologies developed for the base of the economic pyramid. This book is intended for students, faculty, entrepreneurs, and design practitioners working in the areas of appropriate design, rural development, or low-resource and frugal innovation. Case studies presented in the book also provide practical insights for development practitioners, NGOs, policymakers, and engineers.

Organisation of the Book

Chapter 1 introduces the concepts of the base of the pyramid and low-resource settings. It also explains the role of technology in poverty alleviation. Different concepts of decision-making, design methods, and ergonomics are also introduced.

In Chap. 2, technologies developed specifically for the Base of the Pyramid are explained. The technologies discussed in this chapter are chosen from some of the ongoing and completed projects of the Rural Technology Action Group (RuTAG) at the Indian Institute of Technology (IIT) Delhi.

The selection of appropriate technologies is discussed in Chap. 3. Here, the importance and necessity of selecting appropriate technology are highlighted. Using the design literature and field inputs, technology attributes that affect the BoP users have been identified. Using the Technique for Order of Preference by Similarity to Ideal Solution (TOPSIS) methodology, the technology selection for the three technologies introduced in the previous chapter is demonstrated.

In Chap. 4, structural and dynamic analyses of a few technologies are undertaken. The usage conditions of the BoP environment, along with constraints and limitations pertaining to the BoP context, have been considered in the analyses. For the dynamic analysis, the DeNOC-based multi-body dynamics approach is followed, while ergonomic analysis is done considering the human-machine interaction.

Chapter 5 presents the conclusions and findings of the work. The work is summarised, and the key findings of the research are highlighted. Additionally, recommendations are provided to the designers and practitioners in terms of technology selection, design and analysis, and technology implementation.

Acknowledgements We would like to thank all those who have helped in the preparation of this book. The contributions of the staff, research scholars, students, and interns of the Rural Technology Action Group at IIT Delhi are invaluable. Special thanks to the Office of the Principal Scientific Advisor to the Government of India for their guidance and financial support for the RuTAG project. We would also like to thank the students of the Mechatronics Lab at IIT Delhi for their feedback during various stages of this work.

Without the unwavering support and backing of our families, this would not have been possible. We would like to thank our family members Shwetha Bhat, Shripathi Bhat, Sagar, and Sapna (with Suraj Bhat), Bulu and Esha (with Subir Kumar Saha), and Neha (with Vinay Gupta) for their patience and support during the preparation of this book. The authors express sincere thanks to the Springer team led by Swati Mehershi, who assisted us throughout the process.

Moodabidri, Karnataka, India
Delhi, Delhi, India
Greater Noida, Uttar Pradesh, India
September 2025

Suraj Bhat
Subir Kumar Saha
Vinay Gupta

Contents

About the Authors

Dr. Suraj Bhat is currently an Assistant Professor in the Department of Mechatronics at Mangalore Institute of Technology and Engineering. He obtained his PhD from the Department of Mechanical Engineering, Indian Institute of Technology (IIT) Delhi. His thesis was on the selection and analysis of technologies developed for the base of the economic pyramid. He was also associated with the Rural Technology Action Group (RuTAG), IIT Delhi, where he was involved in several socio-technical projects. His research interests include design for development, decision making, design optimization, and appropriate technology. He has several publications, including journal articles and proceedings in national and international conferences.

Prof. (Dr.) Vinay Gupta is a Professor in the Department of Mechanical Engineering at the IEC College of Engineering and Technology, and currently serves as the director of the institute. He earned his PhD from the Indian Institute of Technology (IIT) Delhi in "Dynamic Model Based Optimum Design of Robotic Systems" from the Mechanical Engineering Department. He has more than two decades of teaching and research experience. He has led several projects as Principal Investigator in collaboration with the Rural Technology Action Group (RuTAG) at IIT Delhi. Prof. Gupta co-supervised a PhD candidate at IIT Delhi, focusing on technology selection, analysis, and design for rural and economically weaker sections. He is the co-author of the book titled *Dynamics and Balancing of Multibody Systems: Applications to Rural and Robotic Systems*, published by Springer Nature in 2025. His research interests include multibody dynamics, mechanisms, optimum design, and robotics.

Prof. Subir Kumar Saha is a Professor in the Department of Mechanical Engineering, Indian Institute of Technology (IIT) Delhi since 1996 and currently serves as the Project Director of the I-Hub Foundation for Cobotics at IIT Delhi (since 2020). He holds a BE in Mechanical Engineering from RE College (now NIT), Durgapur (1983), MTech from IIT Kharagpur, and a PhD from McGill University, Canada (1991). Previously, he worked as a Research Scientist at Toshiba

Corporation's R&D Center in Japan (1991–1995) and has been a visiting faculty at IIT Madras, McGill University (Canada), Monash University (Australia), and the University of Verona (Italy). His accolades include the Humboldt Fellowship (1999) by the AvH Foundation, Germany, the Distinguished Alumnus Award from NIT Durgapur (2020), and recognition by his high-school alma mater Midnapore College (2024). Prof. Saha's research interests focus on robotics, multibody dynamics, rural technology applications, and innovative teaching and research methodologies. He established the Mechatronics Laboratory at IIT Delhi in 2001 and contributed to establish the Programme for Autonomous Robotics (PAR) Laboratory in 2010. He has authored textbook, several specialized books, peer-reviewed journal and conference publications, and completed many national- and international-level sponsored projects and consultancies.

Chapter 1
Introduction

In the year 2015, the United Nations held its 70th General Assembly. As a part of this, the United Nations Sustainable Development Summit was also held in New York. One of the important outcomes of this summit was setting the development agenda for 2030, known as the Sustainable Development Goals (United Nations 2015). More than 150 world leaders came together and pledged to implement these goals in the next 15 years. As can be deemed obvious, out of the 17 sustainable goals that were announced, the first one was to end poverty. The goal was to eradicate extreme poverty from the world by 2030. Even though the threshold income limit to categorise someone as poor may vary from region to region, the International Poverty Line is considered a standard across the world. As of 2020, extreme poverty is defined by The World Bank as people living under US$1.90 per day (World Bank Group 2020). This amounts to approximately ₹152 as per the exchange rate of ₹80 per US dollar.

In 2015, 10% of the world's population lived in extreme poverty. In 2019, that percentage had decreased to 8%, which comes out to be around 590 million people. Even though there has been a substantial decrease, the number of people living in poverty is still quite large. Apart from financial implications, living in extreme poverty leads to the person not having access to basic facilities such as food, shelter, and healthcare. As was seen during the COVID-19 outbreak, the poor were more impacted due to the unavailability of resources rather than the virus itself. The World Bank has also indicated that, due to COVID-19 alone, poverty has risen for the first time since 1998 (UNDP 2022). Those in the *extremely poor* category fall into what is known as 'the poverty trap', from which it is very difficult to escape.

It is often assumed that extreme poverty is limited to the rural populace. However, it can be seen that extreme poverty exists among both the rural and the urban populations. One of the reasons is the disproportionate migration of people in search of livelihood to urban areas. The lack of resources in the rural areas makes the atmosphere less conducive to generating enough livelihoods for the people there. This causes over-dependence on the limited resources available in the urban setup.

S. Bhat et al., *Technologies for the Base of the Economic Pyramid*, Design Science and Innovation, https://doi.org/10.1007/978-981-95-2721-2_1

The traditional sources of livelihood, thus, either become outdated or fade out due to the lack of takers. The blame lies not just on the policymakers for the lackadaisical approach when it comes to rural livelihood but also on the scientific community for not rallying enough resources towards making appropriate technologies for the rural masses.

To address this gap, there is a need for focused innovation. Technologies developed should be affordable, sustainable, and adapted to local contexts. This requires a holistic integration of design, development, and selection of the technologies with the user context in mind.

1.1 Technology and Poverty

Turning the pages of history, one can observe that technology has always been associated with the growth and development of societies. The inventions of breakthrough technologies are seen as important milestones throughout human civilisation. More prominently, historical ages such as the Stone Age, the Bronze Age, and the Iron Age have been named after the technological capabilities displayed by the human race at those particular time periods. As time progressed, the advent of newer and better technologies, aided by the democratisation of scientific knowledge, has led to rapid improvement in the quality of life of the average person.

However, not everyone has reaped the benefits of advanced technologies. The pace of growth has not been equal for all. The ever-increasing inequality has ensured that the best technologies are beyond the reach of many, resulting in the unending miseries of poverty, debt, and despair.

Polak (2009) observed '90% of the world's designers spend all their time addressing the problems of the richest 10% of the world's customers'. Over the last few decades, various organisations and researchers have tried to address this skew by designing and developing technologies that help in poverty alleviation. While there have been quite a lot of success storeys, the failures of these technologies to achieve the intended results are also high in number. As Ian Smillie mentions in his book Mastering the Machine (Smillie 1991), it is a bit ironic that while we were able to develop technology that could achieve the feat of human landing on the moon in 1969, we have mostly remained unsuccessful, till date, in developing technologies that can effectively tackle humanitarian issues. This can either mean that the humanitarian and social problems are much more complicated or that they have not been getting enough attention from the scientific community.

The importance of technology-aided sustainable development was advocated as early as the 1920s when Mahatma Gandhi campaigned for contextualising Science and Technology (Gandhi 1929; Ninan 2009). Post-World War II, through campaigns such as the Appropriate Technology movement, there have been efforts to build technologies that were more suited to the local needs and contexts. It was realised that simply transferring a technology from one place to another does not guarantee its success. Throughout this process, there have also been discussions on the definitions of technology. What was earlier seen as a standalone product has now gotten its scope broadened to include social, cultural, and political dimensions.

1.2 Base of the Pyramid and Low-Resource Settings

Prahalad and Hart, in the year 2002, observed that Multinational Corporations (MNCs) had the potential to lift billions of people out of poverty through meaningful investments. The authors also pointed out that, in order to compete at the bottom of the economic pyramid (Fig. 1.1), radical technology innovations and new business models were required (Prahalad and Hart 2002). Originally coined in the context of the business lexicon, the concept of 'Bottom (Base) of the (economic) Pyramid (BoP)' has garnered considerable interest within academia and the scientific community. However, the unique set of socio-technical constraints based on user behaviour and requirements has made it challenging for the researchers to develop plausible solutions that can suitably address the needs of people at the BoP.

In fields such as healthcare, it was observed that much state-of-the-art healthcare equipment imported into developing nations was lying unused due to the lack of training and support. Some of the root causes identified include the failure on the part of the designers to properly understand the context and requirements of the users (Aranda-Jan et al. 2016). The limitations of remote design and

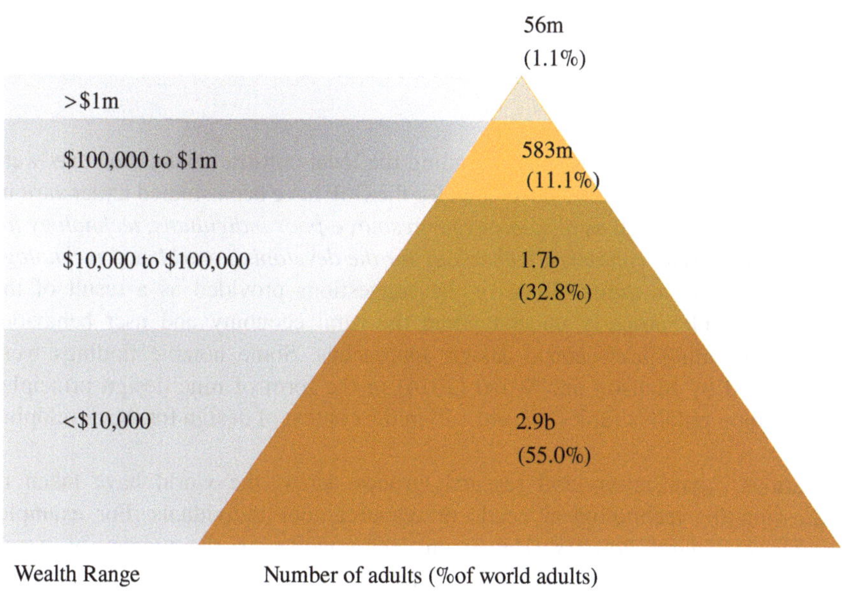

Fig. 1.1 Global wealth pyramid (Shorrocks et al. 2021)

Table 1.1 Design principles for the developing world (Mattson and Wood 2014)

Design principles
1. Co-design with people from the specific developing world context encourages designer empathy, promotes user ownership, and empowers resource-poor individuals
2. Testing the product in the actual setting is an essential part of design for the developing world, not merely a final step
3. Importing technology without adapting it to the specific developing world context is ineffective and unsustainable
4. Both individuals in urban and rural contexts can benefit from poverty alleviation efforts
5. Women and children are more affected by poverty alleviation efforts than men
6. Project management techniques that are adapted to the specific developing world context enable a more effective and efficient design process
7. Products for the developing world have a greater impact when contextualised, developed, and implemented by interdisciplinary teams
8. Cooperation with governments and local influencers contextualises and enables poverty alleviation plans
9. There are existing distribution strategies that can be used to successfully introduce products into developing world markets

Table 1.2 Common pitfalls while designing (Wood and Mattson 2016)

Common pitfalls
1. Lacking the contextual knowledge needed for significant impact
2. Neglecting to make a plan for or developing partners for long-term sustainability
3. Assuming the needs of the customers being served
4. Not making a plan for or developing partners for manufacturing
5. Lacking skills or expertise for a specific project
6. Miscommunicating or failing to develop trust with local stakeholders
7. Forgetting that communities change over time between field visits

technology transfer without understanding the local constraints and contexts were also observed. Technologies developed for the BoP have been studied under various research headings such as *technology for resource-poor individuals*, *technology for less industrialised economies*, *technology for the developing world*, and *technology for low-resource settings*. Some of the suggestions provided as a result of the findings include properly understanding the rural economy and user behaviour and incorporating user-centric design approaches. Some notable findings were documented by Mattson and Wood (2014) in the form of nine design principles and common pitfalls (Tables 1.1 and 1.2) in the context of design for the developing world.

Various organisations and research groups across the world have taken to addressing the technological needs of resource-poor individuals. For example, *Engineers Without Borders* (Helgesson 2006) works on the themes of water, sanitation, energy, waste, and sustainability. The group is present in 26 countries across the globe. International Development Enterprises (IDE) is another non-profit organisation that promotes poverty alleviation through entrepreneurship. In India, groups such as the Rural Technology Action Group (RuTAG) (RuTAG IIT Delhi 2017; Gupta et al. 2019) have leveraged the expertise available in academia to carry out demand-driven technological interventions for the BoP population.

1.3 Technology Selection

Selecting the appropriate technology is a crucial step for any development practitioner. It can be observed, especially in the case of BoP, that the user requirements and constraints vary a lot from region to region. This implies that a technology that has worked for one particular community and context does not necessarily work for another. Technology selection can be viewed as a decision-making problem. In the context of BoP, where multiple factors are to be considered, it is treated as a problem of Multiple-Criteria Decision-Making (MCDM). MCDM can be categorised as Multiple Attribute Decision-Making (MADM) and Multiple-Objective Decision-Making (MODM) (Hwang and Yoon 1981). MODM is more useful in designing, wherein conflicting objectives are considered. Here, the alternatives are not predefined, whereas, in MADM, the best among the alternatives is selected by comparing them on the basis of various attributes. Complex decision-making problems can be represented and solved in a structured and logical manner using MADM techniques.

Several methods, such as Analytical Hierarchy Process (AHP), Technique for Order Preference by Similarity to Ideal Solution (TOPSIS), and Elimination et Choix Traduisant la Realite (ELECTRE), were used for such decision-making in various fields of applications such as management, design, engineering, and economics (Agrawal et al. 1991; Bhangale et al. 2004; Kalbar et al. 2012; Behzadian et al. 2012). In many methods, weights were assigned to the attributes, and the alternatives were ranked based on their overall scores. Attribute weights are generally assigned based on the discretion of the decision-makers or from user experience. Some methods also take care of ratings from multiple decision-makers, where weights are also assigned to each decision-maker based on their importance.

Identification of attributes is an important process before carrying out the comparison and ranking. The results of the MADM methods are only as good as the attributes considered. The chosen attributes must be not only relevant to the user but also appropriate to the context. In the case of BoP technologies, it is essential that, in addition to the technology's performance, attributes that affect the user financially and socially must also be considered. Further, the weights of the attributes can be assigned based on user interviews, field observations, and the experience of the decision-makers.

1.4 Design, Analysis, and Optimisation

There are multiple definitions of design. In general, *designing* can be defined as bridging the gap between the desired situation and the existing situation (Dieter et al. 2009; Simon 2019; Cross 2021). The engineering design process (Fig. 1.2) begins with *need analysis*, followed by *problem definition*, *design synthesis*, *analysis and optimisation*, *evaluation*, and *presentation* (Shigley 2011). The first two stages define the direction of the design process and hence are considered to be the crucial phases of design. In the case of BoP, the stages of *need analysis* and *problem definition* have to be carried out with the help of field research and user surveys. Due to the resource constraints and specific requirements, it becomes important that the designer clearly establishes the context of the design before proceeding to the next phase.

It is to be noted that many technologies being used by the BoP are traditional. The design iterations and changes are often a result of local needs. The design, thus evolved, usually does not have any scientific backing or validation. As a result, many users face difficulties related to productivity and health. Some of the factors, such as structural stability and ergonomics, that are otherwise considered in the design process are usually not taken care of in the case of these technologies. Existing technologies being used by the BoP can be studied and analysed for their overall appropriateness using various principles of mechanical design.

Devices manufactured in a BoP setting generally do not have the required tolerances. Since the manufacturing processes and the tools used do not match high industrial standards, achieving the required tolerances is difficult. As a result, larger clearances are observed in the joints of mechanisms, which leads to performance degradation over time. This also results in the loss of time and money for the users. Furthermore, most of the BoP technologies are the primary sources of livelihood for the users, and they tend to spend most of their time working with these devices. Hence, ergonomic considerations become an important part of the design. However, it can be seen that due to improper ergonomics, users experience various health issues, such as musculoskeletal disorders. Using the concepts of multi-body dynamics and biomechanical modelling of the human body (Fig. 1.3), ergonomic analysis can be carried out. Considering the human-device interactions, reactions, and torque can be calculated at various joints of the human body. Further, using

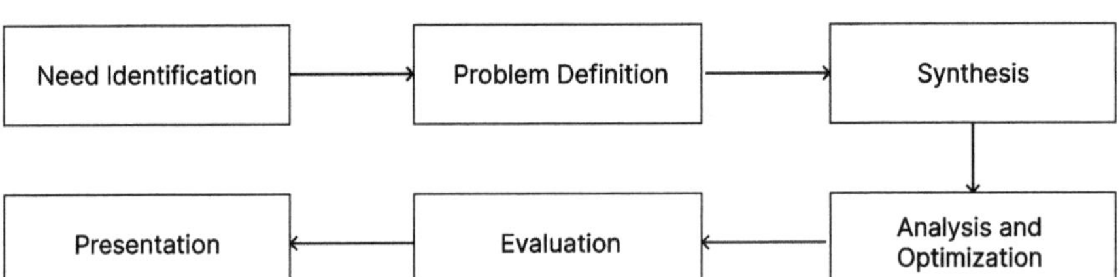

Fig. 1.2 Phases in design (Shigley 2011)

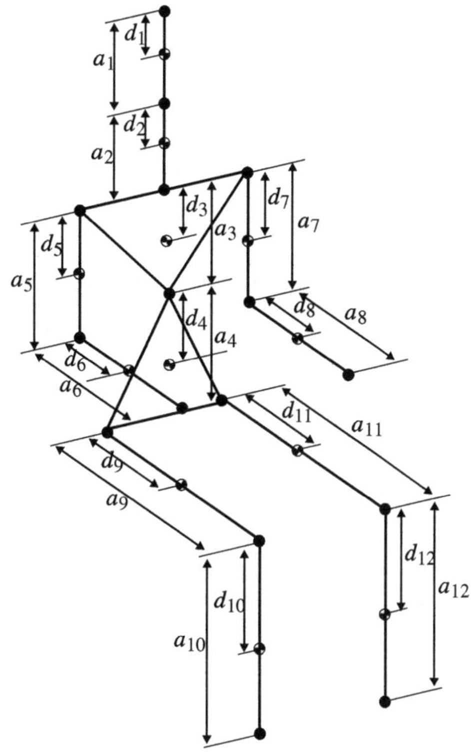

Fig. 1.3 Multi-body model of a human body (Silva et al. 1997)

the concept of muscle fatigue, one can estimate the maximum duration the user can operate the device comfortably.

Based on the results of the analysis, optimisation techniques can be used to optimise the designs of the BoP technologies. For instance, mass redistribution and shape optimisation can be done for mechanisms with larger clearances, such that the impact between the journal and bearing is minimised. This can be an effective approach since the material and labour costs in the BoP settings are lower than the cost of precision machining required to achieve tight tolerances. Similarly, based on the results of the ergonomic analysis, design configuration and dimensions can be optimised to maximise the duration the users can operate the device without compromising on the output.

References

Agrawal VP, Kohli V, Gupta S (1991) Computer aided robot selection: the 'multiple attribute decision making' approach. Int J Prod Res 29(8):1629–1644. https://doi.org/10.1080/00207549108948036

Aranda-Jan CB, Jagtap S, Moultrie J (2016) Towards a framework for holistic contextual design for low-resource settings. Int J Design 10(3):43–63. https://doi.org/10.17863/CAM.7254

Behzadian M, Khanmohammadi Otaghsara S, Yazdani M, Ignatius J (2012) A state-of the-art survey of TOPSIS applications. Expert Syst Appl 39(17):13051–13069. https://doi.org/10.1016/j.eswa.2012.05.056

Bhangale PP, Agrawal VP, Saha SK (2004) Attribute based specification, comparison and selection of a robot. Mech Mach Theory 39(12 SPEC. ISS.):1345–1366. https://doi.org/10.1016/j.mechmachtheory.2004.05.020

Cross N (2021) Engineering design methods: strategies for product design. John Wiley & Sons, London

Dieter GE, Schmidt LC et al (2009) Engineering design, vol 4. McGraw-Hill Higher Education, Boston

Gandhi MK (1929) Urban v. rural. Young India, pp 244–245

Gupta RK, Singh DP, Sharma M, Chatterjee S, Saha SK (2019) Improved technologies under the rural technology action group (RuTAG) at IIT Delhi. In: Rural technology development and delivery. Springer, Berlin, pp 31–45

Helgesson CI (2006) Engineers without borders and their role in humanitarian relief. IEEE Eng Med Biol Mag 25(3):32–35

Hwang CL, Yoon K (1981) Multiple attribute decision making. Lecture notes in economics and mathematical systems, vol 186. Springer, Berlin/Heidelberg. https://doi.org/10.1007/978-3-642-48318-9

Kalbar PP, Karmakar S, Asolekar SR (2012) Selection of an appropriate wastewater treatment technology: a scenario-based multiple-attribute decision-making approach. J Environ Manag 113:158–169. https://doi.org/10.1016/j.jenvman.2012.08.025

Mattson CA, Wood AE (2014) Nine principles for design for the developing world as derived from the engineering literature. J Mech Design 136(12):121403. https://doi.org/10.1115/1.4027984

Ninan AS (2009) Technology as themes of politics. Sustain Development 196(November 2008):183–196

Polak P (2009) Out of poverty: what works when traditional approaches fail. Berrett-Koehler Publishers

Prahalad CK, Hart SL (2002) The fortune at the bottom of the pyramid. Strategy Business (26). https://www.strategy-business.com/article/11518?gko=9b3b4

RuTAG IIT Delhi (2017) Rural technology action group. http://rutag.iitd.ac.in/

Shigley JE (2011) Shigley's mechanical engineering design. Tata McGraw-Hill Education

Shorrocks A, Davies J, Lluberas R (2021) Global wealth distribution 2020. In: Global wealth report 2021. Credit Suisse Research Institute, pp 17 – 25. https://www.credit-suisse.com/about-us/en/reports-research/global-wealth-report.html

Silva MT, Ambrósio J, Pereira MS (1997) Biomechanical model with joint resistance for impact simulation. Multibody Syst Dyn 1(3):65–84. https://doi.org/https://doi.org/10.1023/A:1009700405340

Simon HA (2019) The Sciences of the Artificial, reissue of the third edition with a new introduction by John Laird. MIT Press, Cambridge

Smillie I (1991) Mastering the machine: poverty, aid and technology. Intermediate Technology Publications

UNDP (2022) United nations development programme annual report 2021

United Nations (2015) The 17 goals | sustainable development. https://sdgs.un.org/goals

Wood AE, Mattson CA (2016) Design for the developing world: common pitfalls and how to avoid them. J Mech Design 138(3):031101. https://doi.org/10.1115/1.4032195

World Bank Group (2020) Poverty and shared prosperity 2020

Chapter 2
Technologies for the BoP

In this chapter, we look into some of the works done in the field of technology design and development for the Base of the Pyramid (BoP). A few technologies developed by the Rural Technology Action Group (RuTAG) of IIT Delhi for the BoP have also been introduced. They are projects undertaken as a part of demand-driven technology interventions in rural areas.

2.1 Design for the Developing World

The failure of technology aids sent by the industrialised nations to the developing *'Third World'* countries resulted in the *Appropriate Technology* movement. Technology transfer was found to be ineffective as, over time, resources were not available in the developing countries to carry out maintenance and repair of the imported technologies. Moreover, there was a mismatch between the context in which the technology was developed and the intended beneficiaries in terms of social, political, and cultural considerations. The issues of external dependencies and passive responses with respect to the transferred technologies were observed by the researchers (Teitel 1978; Wood 2017). It was also found that there was a lack of long-term support for the transferred technologies. The Appropriate Technology movement tried to contextualise the technological solutions.

In India, Gandhi was one of the first proponents of the idea of appropriate technology. He underlined the importance of village industries and how 'production by the masses' is more appropriate to the Indian context than the labour-saving 'mass production' techniques (Gandhi 1929; Ninan 2009). Later on, the idea grew into a formal movement of Appropriate Technology across the world with the works of the likes of Schumacher and Papanek (Schumacher 2011; Papanek and Fuller 1972). Various researchers have defined Appropriate Technology (AT) over the years. For instance, Date (1984) defines AT as 'a set of ideas or framework within

S. Bhat et al., *Technologies for the Base of the Economic Pyramid*, Design Science and Innovation, https://doi.org/10.1007/978-981-95-2721-2_2

which to think and act for the development of a society'. Reddy (1979) widened the scope to include societal self-reliance and environmental sustainability. Prahalad (2005), Fisher (2007), and Polak (2009) advocated that development through technologies is not sustainable unless profits are made. Prahalad (2005) coined the term 'Bottom of the Pyramid' for the people living with limited resources. He observed that there was a huge untapped opportunity to create and market goods for the BoP. Fisher (2006) explained how community-owned technology aids failed to achieve any of the intended impacts. He further states that the concept of 'giving away' creates inappropriate incentives leading to patronage and dependencies, which is antithetical to the idea of development.

As a result of the above experiences, researchers, practitioners, and designers came up with various design philosophies and frameworks to design solutions effectively for the BoP context. It was observed that the existing design practices inherently promoted product-centricity. Researchers found that product-centricity limited the scope of the design process and thereby made it less effective while designing for the BoP (Ceschin and Gaziulusoy 2016; Khadilkar 2017a). It was also observed that product-centricity was one of the important factors for the failure of many well-funded projects such as '*One Laptop per Child*' and '*Playpump Water System*' (Khadilkar 2017b). To overcome product-centricity, various user-centric design approaches were proposed. Some of them are discussed below.

2.1.1 Co-design and Co-creation

Co-design encourages collaborations with the end users throughout the phases of design (Mattson and Wood 2014). This includes the stages of need identification, problem formulation, brainstorming, prototyping, and field testing. It can be seen that involving users in the solution process right from the beginning increases the solution's acceptability due to a sense of ownership among the users (Mattson and Wood 2014). Moreover, it increases the efficacy of the design process because of multiple interactions between the design team and the users (Diehl and Christiaans 2007; Nieusma and Riley 2010).

2.1.2 Human-Centred Design (HCD)

HCD is a set of design practices aimed at reducing product-centricity by involving users to understand their needs better. HCD includes the concepts of co-design, participatory design, and empathy design. One of the notable tools used to follow the HCD approach is the IDEO field guide for human-centred design (IDEO 2015), which guides the design team in the various processes and activities involved in the HCD approach.

2.1.3 Design Thinking

Design thinking has a wide range of applications. It also emphasises the use of a human-centred approach for applications in product design and business (Brown and Wyatt 2010). Brown explains how a design process can be simplified into three phases, namely inspiration, ideation, and implementation. He opines that, instead of simply designing a product, it is important to conceive a fully developed market and the ecosystem around it. Practitioners have also applied the concepts of design thinking to create solutions for various developmental projects (Ranger and Mantzavinou 2018).

2.2 Sustainable Development

Sustainability is also an important aspect of *design for development*. Researchers have tried to evaluate existing technologies and innovations in terms of sustainability. The number has only increased after the United Nations formulated and implemented the Sustainable Development Goals (SDGs). For instance, Patnaik and Bhowmick (2018, 2019) explored the *sustainability* dimension of appropriate technologies. The authors speak of how technology should not be short-lived and how it should enable continuous and sustained development. The authors also look into the various constraints in implementing the same. Kalbar et al. (2012) recommended the inclusion of *sustainability* as a criterion for decision-making in selecting appropriate technologies. A framework to reorient science and technology innovations towards SDGs was proposed by Walsh et al. (2020). Technologies, especially those related to energy, were also evaluated based on their contributions towards achieving SDGs (AlQattan et al. 2018; Lohani et al. 2021; Elavarasan et al. 2021).

2.3 RuTAG and Its Design Approach

RuTAGs, or Rural Technology Action Groups in India, were set up by the office of the Principal Scientific Adviser (PSA) to the Government of India. It was observed that while India is on par with the global leaders in some technologies, such as nuclear and space, India is also in a position to leapfrog into global leadership in some areas, such as the rural development tools (Chidambaram 2007). However, the efforts made on rural technology development, especially in the non-farm sectors, did not result in any significant impact. RuTAGs were created as one of the initiatives to address these deficiencies. RuTAG centres are located in the premier technical institutes of the country. As of 2022, RuTAG centres are present in seven

Indian Institutes of Technology (IIT). They are run by teams of faculty, researchers, staff, and student volunteers.

RuTAG centres carry out demand-driven technological interventions to address the shortcomings in the existing rural technologies. The problems are identified by conducting regional workshops with the help of local NGOs or government bodies. The shortlisted problems are then floated as projects so that the faculty and staff of the institutes can get involved. Outwardly, the problems usually look like straightforward engineering problems. However, complexities arise when one considers the socio-cultural aspects of the problem statements. In such cases, it is observed that the conventional design approach often fails to create any impact. RuTAG follows human-centric design approaches as discussed in Sect. 2.1. Various tools of design thinking approaches, such as *participatory rural appraisals*, field visits for need identification and validation, brainstorming, co-design, and co-creation, are used by RuTAG centres. A user-centric problem-solving approach is followed, where the user is involved right from the stage of problem identification to prototyping and testing. Once the solution has been developed, exhibitions and awareness sessions are also undertaken for the propagation of the technologies. Some of the technologies that have been or are being improved by RuTAG IIT Delhi have been explained in the following sections.

2.4 Treadle Pump

A treadle pump is a manual irrigation pump that can draw water from up to a depth of 7–10 metres using atmospheric pressure. They are operated by pedalling the levers alternately using feet. Treadle pumps are primarily used in developing countries and places where groundwater is within the accessible limit.

2.4.1 History

The treadle pump was invented by two development workers, Gunnar Barnes and Dan Jenkins, in the year 1980. Barnes was a Norwegian technologist working with the Rangpur-Dinajpur Rural Service (RDRS), and Jenkins was an engineer working with the United States Agency for International Development (USAID). The high water table throughout Bangladesh and the easily available, cheap manual labour made the technology appropriate to that context. The conception of the idea of the treadle pump is explained in Barnes (1986). The original design of the treadle pump (Fig. 2.1) consisted of two cylinders made of steel sheets. This suction pump was then connected to a tubewell made of bamboo or PVC pipes. It also consisted of a two-plate piston along with a steel piston rod, which was connected to bamboo treadles using a hook and pin joint. The operator could pump out water by treadling on the bamboo levers. The original design also consisted of a rope and pulley arrangement connected to the levers so that by pressing one of the levers, the second

Fig. 2.1 Original design of the treadle pump (Barnes 1981)

one is lifted up. Over time, there have been design changes mostly by the users themselves or, in some cases, by designers. The two-plate pistons were replaced with the No. 6 piston (RWSN 2013), which were the most common pistons being used in hand pumps. The No. 6 pistons consisted of valves and seals made of PVC and were called so because of their use in the 'Number 6' drinking water hand pumps. The rope-pulley mechanism was replaced by a *dheki*-type treadle pump, where the user could pump water by shifting their weight sideways (Kroeker 1989). In the *dheki* treadle pump, the treadles are pivoted onto the main frame, and the weight balance is done in such a way that the treadles lift themselves due to the unbalanced mass on the opposite side. The unbalanced mass meant that the user had to apply additional force to overcome it. It should be noted how the users have given priority to comfort and ease of use over efficiency and performance. The pump was noticed by several non-government and government organisations, which promoted and popularised the treadle pump further. RDRS initially manufactured treadle pumps in their workshops. Later, organisations such as the International Development Enterprises (IDE) in Bangladesh and the International Rice Research Institute (IRRI) in the Philippines commercialised and disseminated the technology in their respective areas and abroad. In Africa, organisations such as IDE, Kickstart (previously known as ApproTEC), and Enterprise Works have been successful in manufacturing and disseminating treadle pumps.

2.4.2 *Classification of Treadle Pumps*

There have been different types of treadle pumps designed and used. The main principle, however, remains the same. The objective is to generate reciprocating motion for the piston to traverse and generate suction pressure. Based on the data collected from the literature (Kay and Brabben 2000; Bhat et al. 2020), treadle pumps are classified based on discharge type and treadling mechanism.

2.4.2.1 Discharge Type

Based on the discharge type, treadle pumps can be classified as suction pump or pressure pump.

i. Suction Pumps

Suction pumps are the original treadle pumps designed in Bangladesh. These pumps lift water from a source not deeper than 7 m–10 m and deliver it at atmospheric pressure through a channel. The water is then routed to irrigate the farms using small canals.

ii. Pressure Pumps

The needs of African farmers varied from those of Bangladeshi farmers. Instead of irrigating the farms using canals, they preferred sprinklers, drips, and hoses (Kay and Brabben 2000). The Bangladeshi treadle pumps were modified at the delivery end so that the water could be delivered through pipes under pressure that could then be fed to sprinklers or overhead tanks.

2.4.2.2 Treadling Mechanism

Based on the treadle mechanism, four different types of treadle pumps have been identified.

i. With Rope and Pulley

This is one of the initial designs of the treadle pump. As shown in Fig. 2.2a, long bamboo or metal levers are used as treadles. The fulcrum is located at the far end of the lever. This is a class II lever, where the output force is between the input force and the fulcrum. The pistons are attached to the levers using pins. A central frame is present, to which a pulley is attached. The levers are connected to the pulley by ropes. This arrangement maintains the reciprocating motion of the piston in such a way that when one lever is pressed, the corresponding piston goes down, and the other piston rises up along with the second lever. Commercial models also include some variations of the original model, such as the use of metal levers and a chain-rocker arrangement.

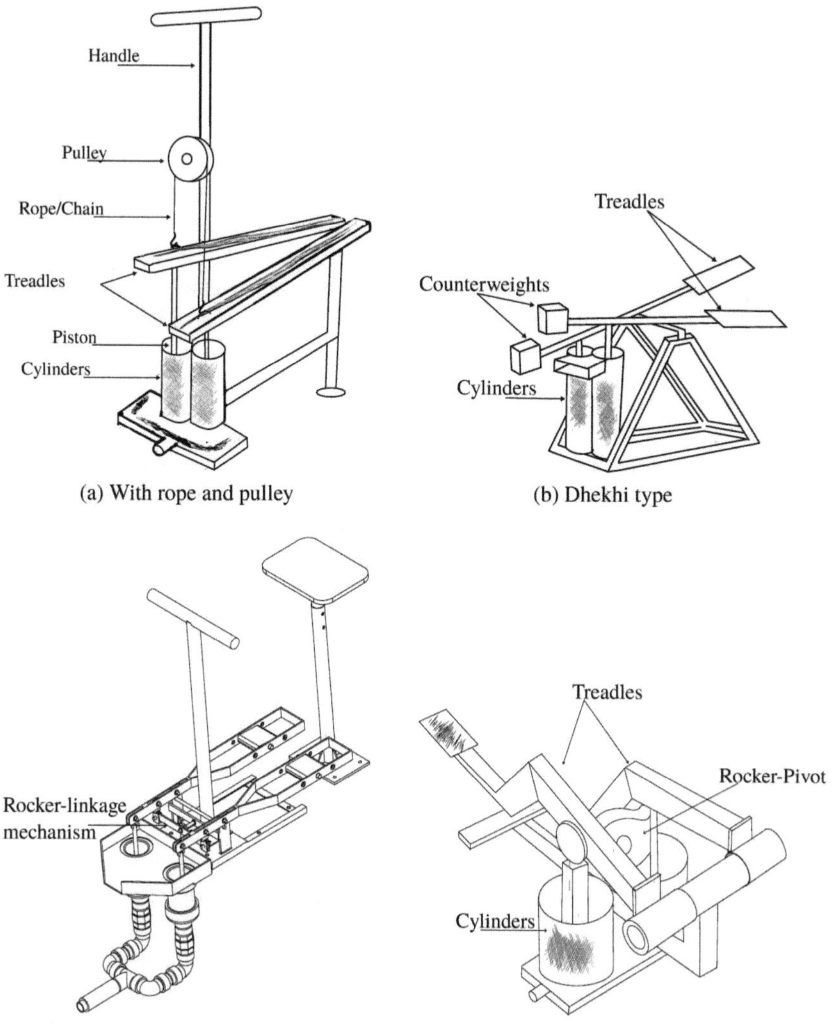

Fig. 2.2 Types of treadle pumps. (**a**) With rope and pulley. (**b**) Dhekhi type. (**c**) Rocker linkage. (**d**) Rocker pivot

ii. *Dheki* Type

Dheki-type treadle pump model (Fig. 2.2b) was one of the initial variations from the original design. This is a class I lever with a fulcrum between the load and effort. In this model, the treadles are pivoted to the main frame and are not coupled with each other. The levers return to the neutral position because of the weight distribution or counterweights attached to the levers. Bamboo levers with appropriate weight distribution or metal levers with counterweights are used in commercial models of this type.

iii. With Rocker Linkage

This is an improvement over the *dheki* type wherein, instead of counterweights, a rocker linkage (Fig. 2.2c) is used to achieve reciprocation. A four-bar mechanism is added to couple the two levers. However, unlike the rocker-chain arrangement, the coupling mechanism is situated below the levers. This is a class I lever mechanism.

iv. With Rocker Pivot

In this type of treadle pump, a rocker pivot (Fig. 2.2d) situated below the levers is used to achieve the reciprocating motion. Class II levers are used here. The rocker pivot performs the same function as the rocker lever mechanism by using a single rigid pivot instead of the three additional linkages.

2.5 *Tulsi Mala* Bead-Making Device

Tulsi malas are garlands of *Tulsi* (Holy Basil) beads are used in rituals such as chanting and prayers. Over 2000 families residing in the *Brij* areas near Mathura, Vrindavan, and Bharatpur are involved in the bead-making process. Beads are made from the stems of *Tulsi* plants cultivated in the nearby regions. Traditionally, beads were made using a hand-operated lathe-like device. However, the recent interventions by RuTAG at IIT Delhi (Gupta et al. 2015, 2019; Prasad and Saha 2022) and local innovations from the field have resulted in the development of multiple mechanised *Tulsi mala* bead-making devices. These innovations have risen out of the varying needs across the artisan groups. It is to be noted that various sizes of beads, ranging from 2 mm to 15 mm in diameter, can be made using these devices.

2.5.1 Various Models

Various types of *Tulsi mala* bead-making devices are currently available and are being used by artisans. The devices range from traditional hand-operated devices to pedal-operated AC-powered devices. The basic operation principles of the device are turning and drilling. Thus, it can be considered a miniature lathe. Some of the typical components of the bead-making device include a tool post, a pin for making a hole in the beads, a stem holder, and a base frame. The available devices are discussed below.

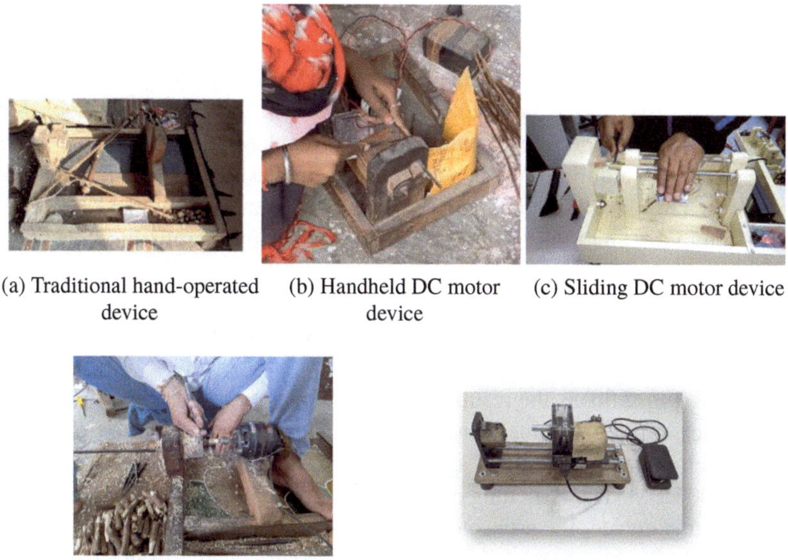

(a) Traditional hand-operated device (b) Handheld DC motor device (c) Sliding DC motor device

(d) Improvised traditional device with AC motor (e) Pedal-operated AC motor powered device

Fig. 2.3 Various *Tulsi mala* bead-making devices. (**a**) Traditional hand-operated device. (**b**) Handheld DC motor device. (**c**) Sliding DC motor device. (**d**) Improvised traditional device with AC motor. (**e**) Pedal-operated AC motor powered device

2.5.1.1 Traditional Hand-Operated Device

These are the traditional devices (see Fig. 2.3a) used by the artisans over the years. Here, the rotation is provided by manually cranking the stem holder. *Tulsi* stems are attached to the stem holder. It is to be observed that both hands are engaged while operating this device. While cranking is done with the left hand, the turning of the bead is done with the right hand. The artisan has to sit on the floor while operating the device. The artisans can produce beads ranging from 2 mm to 15 mm in diameter using this device. However, it is highly dependent on the skill and experience of the artisan.

2.5.1.2 Handheld DC Motor Device

In this device (Fig. 2.3b), a 12-V DC motor provides rotation to the *Tulsi* stem. The motor is held by the artisan in one hand, while with the other hand, the bead is cut. These devices are often used to make small beads of size 2 mm–3 mm in diameter, which are used to make necklaces, also known as *Kanthi mala*. Here, ball pen refill tubes are used as stem holders. Generally, these *malas* are made by women, and their cost varies from ₹90 to ₹120 per 20 *malas*.

2.5.1.3 Sliding DC Motor Device

The improved version of the handheld DC motor device was developed by RuTAG IIT Delhi in 2012. This device (Fig. 2.3c) consists of supporting rods for the movement of the motor, a limit switch, a speed regulator, and an adapter. The stem holder is made of aluminium with a tapered square groove. This device can produce beads from 4 mm to 10 mm in diameter. These beads are used to make *kanthi malas* (4 mm–5 mm dia.) and *japa mala* for meditation (6 mm–10 mm dia.).

2.5.1.4 Improvised Traditional Device with AC Motor

This device (Fig. 2.3d) is the result of a field innovation where the artisan has fixed an AC motor in the traditional hand-operated device. The motor rests on a wooden block, which is moved using the artisan's leg. The operating speed of the motor is around 5000 RPM. The switching of the motor is done by pressing the pedal with the leg. Using this device, the artisan can produce beads of 8 mm–15 mm in diameter. Since the speed is very high, only a skilled and experienced artisan can operate this device.

2.5.1.5 Pedal-Operated AC Motor Powered Device

Based on the feedback received on the sliding DC motor device developed by RuTAG IIT Delhi, a pedal-operated AC-powered device (Fig. 2.3e) was developed in 2021 for making bigger beads. This device can provide enough torque to produce beads of sizes 6 mm to 15 mm in diameter. The motor slides over guide rails supported by linear bearings. This device can be operated while sitting either on the floor or on a chair, as per the convenience of the artisan. The operating speed of this device is around 2000 RPM.

2.5.2 Type of Beads

Using the available devices, artisans produce different sizes of beads (Fig. 2.4) used for different purposes. Four different sizes of beads are described below.

2.5.2.1 *Kanthi Mala* Beads

Kanthi mala beads are the smallest beads used to make necklaces or *kanthi malas*. Their sizes vary from 2 mm to 5 mm in diameter. The stems to produce these beads are purchased at around ₹30 to ₹40 per kg, and a set of 20 *malas* is sold at ₹90 to ₹120.

Fig. 2.4 Various sizes of *Tulsi* beads

2.5.2.2 *Matar* Dana

The next size of beads is called *matar dana* or pea-sized bead. They are of 5 mm to 8 mm in diameter. Stems to produce these are purchased at around ₹90 to ₹120 per kg, and a set of 20 *malas* is sold at ₹400 to ₹600.

2.5.2.3 *Japa Mala* Beads

Japa mala beads are of size 8 mm to 10 mm in diameter. The raw material to make these beads is purchased at around ₹150 to ₹200 per kg. These beads are sold at ₹800 to ₹2000 per set of 20 *malas*.

2.5.2.4 Large Beads

Larger beads of sizes ranging from 10 mm to 15 mm in diameter are also produced. Malas made of these beads are often used as offerings to the deity. Stems to make these beads are purchased at around ₹250 to ₹300 per kg, and a set of 20 *malas* is sold at ₹3000 to ₹6000.

2.6 Street Vending Cart

In many developing countries, street vending is a common source of income (Bhat et al. 2021). Numerous carts offering a variety of goods, including toys, clothing, fresh produce, street food, and more, are often seen on urban streets. Even though they are mostly included in the category of informal employment, street vendors are a vital component of the urban economy. For instance, in Indian cities, street vendors are estimated to make up approximately 2% to 2.5% of the population in metropolitan areas (Naik 2013). The younger generation, who has moved from rural areas in search of employment opportunities, makes up the majority of these vendors. In addition to providing jobs, street vending fulfils a significant public purpose by providing urban consumers with accessible, reasonably priced goods.

Among the different categories of street vending, fruit and vegetable vendors face additional challenges due to the perishable nature of their goods. These items are typically sold from both stationary setups and mobile vending carts. Mobile vending offers the advantage of direct customer access, allowing vendors to move through neighbourhoods rather than compete for space at fixed market locations. However, increased mobility often comes at the cost of reduced load capacity, as heavier carts are more difficult to manoeuvre in dense urban environments.

In much of Asia, Latin America, and Africa, vending carts are a common sight. Depending on local requirements and usage contexts, they differ in form and function. The majority of carts are pushed by hand and are made of metal and wood. But there are also modified versions, like those that are attached to bicycles or fitted with engines. The following section discusses the different types of vending carts currently in use.

2.6.1 *Four-Wheeled Push Cart*

The four-wheeled push cart shown in Fig. 2.5a represents one of the most commonly used and basic types of vending carts. Due to their restricted steering capabilities, these carts are mostly used for stationary vending. The structure is made of wooden planks mounted on a metal frame. The display area generally measures around 1 m × 1.7 m and can carry a load of up to 125–150 kgs. The carts are often customised as per the requirements and sometimes feature wooden boxes to store personal belongings of the vendor.

2.6.2 *Tricycle Pull Cart*

The tricycle-based cart (Fig. 2.5b) is constructed by modifying a conventional tricycle chassis. A wooden platform is mounted onto the chassis for carrying goods. Although the loading area is slightly smaller than that of a conventional four-wheeled push cart, at around 1 m by 1.2 m, the cart offers a similar load capacity of about 150 kg. With modifications, they can carry loads up to 200 kg. This design is particularly suited for vendors who need to travel longer distances. In the tricycle pull cart, the vendor sits in front of the cart.

(a) Four Wheel Push Cart

(b) Tricycle Pull Cart

(c) Tricycle Push Cart

(d) Motorized e-Cart

Fig. 2.5 Types of vending carts. (**a**) Four Wheel Push Cart. (**b**) Tricycle Pull Cart. (**c**) Tricycle Push Cart. (**d**) Motorised e-Cart

2.6.3 Tricycle Push Cart

In this version, the vendor is seated behind the platform while operating the tricycle. An example layout is shown in Fig. 2.5c. The platform dimensions are typically smaller, around 1 m by 1 m, to improve the stability during motion. A common drawback of this design is that the loaded platform may obstruct the vendor's line of sight while cycling.

2.6.4 Motorised e-Cart

The growing adoption of electric vehicles, supported by government incentives, has increased the use of e-rickshaws and e-scooters for commercial vending. Among newer vendors, there is a noticeable preference for motorised carts due to increased comfort and mobility. These vehicles (Fig. 2.5d) are often modified with custom chassis to accommodate a larger volume of goods. Although considerably more costly than conventional manual carts, most vendors consider the investment justified due to the operational benefits.

2.7 Summary

This chapter provided an overview of technologies designed for the Base of the Economic Pyramid (BoP). It began by discussing the characteristics of BoP technologies and how they differ from mainstream ones. These technologies are often simple, affordable, and adapted to the needs of low-income users.

Design challenges in BoP contexts were outlined. These include low resource availability, varied user capabilities, and the need for easy maintenance. The importance of context, usability, and adaptability in the design process was also emphasised.

Three case studies were presented in this chapter. Each of these technologies illustrates how locally adapted, low-cost solutions can support livelihoods and meet essential needs. The treadle pump illustrated how simple mechanical design can support rural irrigation, the *Tulsi mala* bead-making device demonstrated how institutional and local innovations evolve and co-exist, and the street vending cart highlighted the diversity of urban livelihood avenues shaped by various factors. Each case also showed that multiple variants of a single technology may exist. These variants are shaped by user needs, regional practices, and socio-economic contexts. This diversity also makes the task of selecting the most appropriate design variant both necessary and complex.

References

AlQattan N, Acheampong M, Jaward FM, Ertem FC, Vijayakumar N, Bello T (2018) Reviewing the potential of waste-to-energy (WTE) technologies for sustainable development goal (SDG) numbers seven and eleven. Renew Energy Focus 27:97–110

Barnes C (1981) Treadle pump GB drawing. https://commons.wikimedia.org/wiki/File:Treadle_pump_GB_drawing.jpg

Barnes G (1986) The twin-treadle pump: manual irrigation in Bangladesh. In: Small farm equipment for developing countries: proceedings of the international conference on small farm equipment for developing countries: past experiences and future priorities. International Rice Research Institute, p 181

Bhat S, Saha SK, Gupta V (2020) Study and analysis of treadle pump. ISME J Mech Design 4(2):27–35

Bhat S, Saha SK, Gupta V (2021) Attributes based ranking and selection of vending carts using fuzzy TOPSIS. In: 2021 IEEE global humanitarian technology conference (GHTC). IEEE, Piscataway, pp 258–264

Brown T, Wyatt J (2010) Design thinking for social innovation. In: Stanford social innovation review, pp 31–35

Ceschin F, Gaziulusoy I (2016) Evolution of design for sustainability: from product design to design for system innovations and transitions. Design Stud 47:118–163. https://doi.org/10.1016/j.destud.2016.09.002

Chidambaram R (2007) Indian innovation: action on many fronts. Issues Sci Technol 24(1)

Date A (1984) Understanding appropriate technology. In: Appropriate technology in third world development, pp 163–183

Diehl JC, Christiaans H (2007) The first learning experience of designing for the bop. In: IASDR07 Emerging trends in design research. The Hong Kong Polytechnic University, Hong Kong, pp 1–12

Elavarasan RM, Pugazhendhi R, Jamal T, Dyduch J, Arif MT, Kumar NM, Shafiullah G, Chopra SS, Nadarajah M (2021) Envisioning the un sustainable development goals (SDGs) through the lens of energy sustainability (SDG 7) in the post-covid-19 world. Appl Energy 292:116665

Fisher M (2006) Income is development: Kickstart's pumps help Kenyan farmers transition to a cash economy. Innov Technolo Gov Glob 1(1):9–30

Fisher M (2007) Design to kickstart incomes. Design for the other 90

Gandhi MK (1929) Urban v. Rural. In: Young India, pp 244–245

Gupta R, Sharma M, Singh DP, Suthar B, Saha SK (2015) Women empowerment by technology supported manufacturing of beads from holy basil. Curr Sci 109(9):1660–1664

Gupta RK, Singh DP, Sharma M, Chatterjee S, Saha SK (2019) Improved technologies under the rural technology action group (RuTAG) at IIT Delhi. In: Rural technology development and delivery. Springer, Berlin, pp 31–45

IDEO (2015) Design Kit: The Human-Centered Design Toolkit. https://www.ideo.com/post/design-kit

Kalbar PP, Karmakar S, Asolekar SR (2012) Selection of an appropriate wastewater treatment technology: a scenario-based multiple-attribute decision-making approach. J Environ Manag 113:158–169. https://doi.org/10.1016/j.jenvman.2012.08.025

Kay M, Brabben T (2000) Treadle pumps for irrigation in Africa. Tech. rep., Food and Agriculture Organization of the United Nations, Rome

Khadilkar P (2017a) Formulating the design scope for the base of the (economic) pyramid. Design Issues 33(2):4–17. https://doi.org/10.1162/DESI_a_00436

Khadilkar P (2017b) Formulation of a framework for needs analysis and stakeholders' behavioral simulation for design for the bop. PhD thesis, Indian Institute of Science

Kroeker MG (1989) Testing and Analysis of the Bangladeshi Treadle Pump. PhD thesis

Lohani SP, Dhungana B, Horn H, Khatiwada D (2021) Small-scale biogas technology and clean cooking fuel: assessing the potential and links with SDGs in low-income countries—a case study of Nepal. Sustain Energy Technol Assessments 46:101301

Mattson CA, Wood AE (2014) Nine principles for design for the developing world as derived from the engineering literature. J Mech Design 136(12):121403. https://doi.org/10.1115/1.4027984

Naik A (2013) Contextualising urban livelihoods: street vending in India. SSRN Electronic Journal. https://doi.org/10.2139/ssrn.2238589

Nieusma D, Riley D (2010) Designs on development: engineering, globalization, and social justice. Eng Stud 2(1):29–59. https://doi.org/10.1080/19378621003604748

Ninan AS (2009) Technology as themes of politics. Sustain Development 196(November 2008):183–196

Papanek V, Fuller RB (1972) Design for the real world

Patnaik J, Bhowmick B (2018) Appropriate technology and management for sustainability. In: 2018 IEEE international conference on industrial engineering and engineering management (IEEM). IEEE, Piscataway, pp 1289–1292

Patnaik J, Bhowmick B (2019) Revisiting appropriate technology with changing socio-technical landscape in emerging countries. Technol Soc 57:8–19

Polak P (2009) Out of poverty: what works when traditional approaches fail. Berrett-Koehler Publishers

Prahalad CK (2005) The fortune at the bottom of the pyramid: eradicating poverty through profits. Wharton School Publishing

Prasad Y, Saha SK (2022) Development of an AC motor powered device for making Tulsi mala beads. In: Rural technology development and delivery. Springer, Berlin, pp 21–28

Ranger BJ, Mantzavinou A (2018) Design thinking in development engineering education: A case study on creating prosthetic and assistive technologies for the developing world. Development Eng 3(September 2017):166–174. https://doi.org/10.1016/j.deveng.2018.06.001

Reddy AK (1979) Problems in the generation of appropriate technologies. In: Appropriate technologies for third world development. Springer, Berlin, pp 173–189
RWSN (2013) No. 6 pump. https://www.rural-water-supply.net/en/implementation/public-domain-handpumps/no-6-pump
Schumacher EF (2011) Small is beautiful: a study of economics as if people mattered. Random House
Teitel S (1978) On the concept of appropriate technology for less industrialized countries. Technol Forecast Soc Change 11(4):349–369. https://doi.org/10.1016/0040-1625(78)90018-5
Walsh PP, Murphy E, Horan D (2020) The role of science, technology and innovation in the un 2030 agenda. Technol Forecast Soc Change 154:119957
Wood AE (2017) Principles and insights for design for the developing world. PhD thesis, Brigham Young University

Chapter 3
Selection of Appropriate Technologies

Mahatma Gandhi wrote the following in the year 1929 in his weekly journal Young India (Gandhi 1929):

'What is good for one nation situated in one condition is not necessarily good enough for another differently situated. One man's food is often another man's poison. Physical geography of a country has a predominant share in determining its culture. A fur coat may be a necessity for the dweller in the polar regions, it will smother those living in the equatorial regions'.

The debate was on the notion of industrialisation in India, which Gandhi felt came at the price of losing ancient rural traditions. Furthermore, Gandhi opined that in a country with a teeming population, it is necessary that the machinery does not displace labour and thus resulting in the loss of employment. The importance of contextual design was further highlighted when the scientific community realised that most of the technology transferred to the developing nations was lying unused. This triggered the Appropriate Technology movement around the 1970s, where the designers looked for alternative approaches by trying to use local resources and appropriate materials while designing new technologies.

In this chapter, we look at the various approaches to rank and select existing technologies based on the context of their usage. The technologies specifically designed and developed for the BoP markets often follow a different design approach as compared to those developed for the Top of the Pyramid (ToP) (Jagtap et al. 2014). As a result, the solutions thus developed tend to be more localised. The needs and constraints considered for the design process are based on the geographical, socio-cultural, and economic backgrounds of the particular cluster or community. In such cases, it is important to note that the technology designed for one particular context might not be successful in another area. Using Multi-Attribute Decision-Making (MADM), we look at how practitioners can compare and rank technological alternatives. Additionally, we also examine how subjective inputs, such as user preferences and stakeholder opinions, can be incorporated into the selection methodology. This chapter presents three case studies illustrating

S. Bhat et al., *Technologies for the Base of the Economic Pyramid*, Design Science and Innovation, https://doi.org/10.1007/978-981-95-2721-2_3

the application of the selection framework. Each example highlights the user-centric and scenario-driven approaches that reflect the complexity of appropriate technology deployment.

3.1 Technology Selection and Decision-Making

Selecting appropriate technology is a crucial and challenging task. With multiple variants of a single technology available in the market, it becomes difficult for users and practitioners to choose the right alternative for their context. Various researchers have come up with definitions and criteria to judge the appropriateness of technologies. Teitel (1978) writes about the criteria that make a technology appropriate. Based on the results of a survey, the author listed several factors that affect the appropriateness of a technology. Some of them include market, raw materials, scale, labour, machinery, and the cost of technology transfer. Reddy (1979) defined appropriate technologies as those that fulfil the development objectives of satisfying basic human needs, self-reliance, and environmental harmony. Bowonder (1979) has suggested a set of criteria for the evaluation of appropriate technologies. The author talked about how needs are contextual and that preconceived notions regarding the hierarchy of needs defeat the purpose of appropriateness. The author mentioned that appropriate technology has to be selected based on comparison and ranking. He has also listed 20 criteria that can help in the multi-criteria decision-making process. Some of the criteria include energy, labour, cost, raw material, and productivity, among other socio-economic criteria. The importance of selection and the lack of sufficient knowledge available to the decision-makers were also highlighted (Date 1984). Transferring or selecting a technology without considering its appropriateness with respect to the user often results in the failure of the technology (Donaldson 2006; Lissenden et al. 2015; Zhou et al. 2017).

To address these challenges, researchers have turned to Multi-Attribute Decision-Making (MADM) methods for systematic technology selection (Agrawal et al. 1991; Bhangale et al. 2004; Kalbar et al. 2012). These methods help decision-makers evaluate and rank alternatives based on multiple, often conflicting, criteria. One of the foundational classifications of MADM techniques was developed by Hwang and Yoon (1981), who categorised them based on the nature of information available to decision-makers (Table 3.1). Some of the other methods include Multi-Attribute Utility Theory (MAUT) (Huber 1974), Preference Ranking Organisation Method for Enrichment Evaluation (PROMETHEE) (Brans et al. 1986), and Analytical Hierarchy Process (AHP) (Saaty 1977). The methods are also classified based on the decision-making approach (Hwang and Yoon 1981). Models that do not permit trade-offs between attributes are called non-compensatory models. Here, an unfavourable value in an attribute cannot be offset by a favourable value in another attribute. Conjunctive, disjunctive, and lexicography methods are some examples of non-compensatory models. Compensatory models permit trade-offs between attributes, and the outcomes are closer to reality. Some examples of compensatory

Table 3.1 Classification of MADM methods (Hwang and Yoon 1981)

No information	Information about attributes	Information about alternatives
1. Dominance	1. Conjunctive method	1. LINMAP
2. Minimax	2. Disjunctive method	2. Interactive SAW
3. Maximax	3. Lexicography method	3. Multidimensional Scaling
	4. Elimination by aspects	
	5. Permutation method	
	6. Linear assessment	
	7. Simple Additive Weighting (SAW)	
	8. Hierarchical Additive Weighting	
	9. ELECTRE	
	10. TOPSIS	
	11. Hierarchical trade-offs	

models are TOPSIS, ELECTRE, and PROMETHEE. Another classification is based on the type of data: Cardinal and Ordinal. Cardinal data refers to the quantitative data available for the attributes. Ordinal refers to the preference or ranking data. In the case of technology selection, information is usually available on a cardinal scale. The next section provides a brief overview of two popular ranking methods used in this study: AHP and TOPSIS.

3.1.1 Analytical Hierarchy Process (AHP)

AHP uses a pairwise comparison of alternatives based on their relative performance against each attribute (Saaty 1977). The relative preferences are then converted to a set of weights. If a set of n activities $A_1 \dots A_n$ are to be compared according to their relative weights $w_1 \dots w_n$, the pairwise comparison can be represented in the form of a matrix as

$$\mathbf{A} = \begin{array}{c} \\ A_1 \\ A_2 \\ \vdots \\ A_n \end{array} \begin{array}{c} \begin{array}{cccc} A_1 & A_2 & \dots & A_n \end{array} \\ \begin{bmatrix} w_1/w_1 & w_1/w_2 & \dots & w_1/w_n \\ w_2/w_1 & w_2/w_2 & \dots & w_2/w_n \\ \vdots & \vdots & & \vdots \\ w_n/w_1 & w_n/w_2 & \dots & w_n/w_n \end{bmatrix} \end{array} \tag{3.1}$$

If $\mathbf{A}$ is multiplied by the vector $\mathbf{w} = [w_1, w_2, \dots, w_n]^T$, we obtain the vector $n\mathbf{w}$. Hence, it can be seen that

$$\mathbf{A}\mathbf{w} = n\mathbf{w} \tag{3.2}$$

Table 3.2 Saaty's scale for relative importance (Saaty 1977)

Intensity of importance	Definition	Explanation
1	Equal importance	Two activities contribute equally to the objective
3	Weak importance of one over another	Experience and judgement slightly favour one over another
5	Strong importance	Experience and judgement strongly favour one over another
7	Demonstrated importance	An activity is strongly favoured and its dominance is demonstrated in practice
9	Absolute importance	The evidence favouring one activity over another is of the highest possible order of affirmation
2, 4, 6, 8	Intermediate values	When compromise is needed

If the matrix **A** is known, the elements of the vector **w** can then be obtained by solving

$$(\mathbf{A} - n\mathbf{I})\mathbf{w} = \mathbf{0} \tag{3.3}$$

Equation 3.3 has a non-zero solution only if n is an eigenvalue of **A**. It can also be observed that in matrix **A**, every column is a multiple of the first column, and every row is a multiple of the first row. This means that the rank of matrix **A** is unity (Klein 2019). Hence, there is only one non-zero value in the set of eigenvalues $\lambda_i (i = 1, 2, \ldots, n)$, which is equal to the trace of the matrix. Hence, in this case, $\lambda_{max} = n$. To rate the importance, Saaty also proposed a scale that is shown in Table 3.2.
The alternatives are compared by creating pairwise comparison matrices for each activity. The weighted values will then give the overall global priority for each alternative.

3.1.2 *TOPSIS*

Technique for Order Preference by Similarity to Ideal Solution (TOPSIS) is an MADM approach proposed by Hwang and Yoon (1981). This approach uses *similarity index* to rank the alternatives. Unlike AHP, which requires multiple comparison matrices at different hierarchical levels, TOPSIS uses a single decision matrix along with a pairwise comparison matrix of the attributes for ranking.

The decision matrix consists of attribute information of all the alternatives. If m alternatives $(A_1, A_2, \ldots, A_m)$ have n attributes $(a_1, a_2, \ldots, a_n)$, the decision matrix **D** can be written as

$$\mathbf{D} = \begin{array}{c} \\ A_1 \\ A_2 \\ \vdots \\ A_m \end{array} \begin{array}{c} \begin{array}{cccc} a_1 & a_2 & \dots & a_n \end{array} \\ \begin{bmatrix} d_{11} & d_{12} & \dots & d_{1n} \\ d_{21} & d_{22} & \dots & d_{2n} \\ \vdots & \vdots & & \vdots \\ d_{m1} & d_{m2} & \dots & d_{mn} \end{bmatrix} \end{array} \tag{3.4}$$

where d_{ij} is the jth attribute of the ith alternative. Attribute weights are calculated as explained in the AHP method. Further, the decision matrix is normalised to obtain the *normalised decision matrix* **N**, which is then multiplied to the weight vector **w** to obtain the weighted normalised matrix **V**.

TOPSIS uses the concept of *ideal solution* and *separation measures*. Positive and negative ideal solutions, $\mathbf{v}^+$ and $\mathbf{v}^-$, are obtained using the best and the worst values, respectively, in each column. The separation measures are the n-dimensional Euclidean distance of each alternative from the ideal solutions and can be calculated as

$$S_i^+ = \left[\sum_{j=1}^{n} (v_{ij} - v_j^+)^2 \right]^{1/2} \quad (i = 1, 2, \cdots, m) \tag{3.5}$$

and

$$S_i^- = \left[\sum_{j=1}^{n} (v_{ij} - v_j^-)^2 \right]^{1/2} \quad (i = 1, 2, \cdots, m) \tag{3.6}$$

The alternatives are then ranked based on the C^* scores obtained as

$$C_i^* = S_i^- / (S_i^+ + S_i^-) \tag{3.7}$$

3.2 Attribute-Based Selection

The involvement of multiple organisations and researchers around the world has led to the development of multiple solutions for problems related to BoP. This has led to a situation wherein the user or a development practitioner can find it challenging to choose the best alternative for the given context. A wrong choice of technology can often lead to economic, social, and environmental problems (Date 1984). Due to the lack of resources, the preferences of users in the BoP can vary significantly from those at the ToP. For instance, it is observed that the expenditure is prioritised differently, and most of their incomes are spent on food, housing, and household goods (Guesalaga and Marshall 2008). Similarly, with respect to the technology, it is essential that appropriate attributes that affect the users directly are identified.

These attributes act as decision criteria in MADM frameworks, enabling structured comparison and ranking of alternatives. The effectiveness of any decision model depends directly on the relevance and quality of the attributes considered. Therefore, understanding the user's context and identifying the right set of attributes are critical in appropriate technology selection.

3.2.1 User Contexts and Needs

In order to choose the best among a set of alternatives, it is essential that the contexts and needs of the users are adequately understood. To do so, one may look into the design methodologies used to develop new products for the BoP. Literature highlights the differences in the design process for the BoP as compared to the rest and also notes that the needs of users at the BoP are significantly different compared to the needs of other users (Castillo et al. 2012; Jagtap et al. 2014). It was found that the BoP designers place more importance on material, power, and cost requirements. However, on the other hand, it was also noted that simplistic assumptions and focusing solely on the criterion of affordability often led to the failure of the designed products (Kandachar and Halme 2005). Factors such as the need for frequent maintenance, the requirement of additional training, and skill sets also affect the acceptability of the developed products (Andersen and Kim 2011). Design tools such as *Personas* (Cooper and Saffo 1999; Grudin and Pruitt 2002; Rodríguez et al. 2006; Floyd et al. 2008) and *Scenarios* (Bødker 2000) can also be used to extract user needs and visualise the user contexts effectively.

3.2.2 Attribute Identification

In the MADM methods, the alternatives are compared based on various attributes. The rankings thus obtained are only as relevant as the attributes that are considered. Therefore, it is essential that appropriate attributes are identified and considered for the ranking. One of the main distinguishing factors between the BoP market and the rest of the markets is the availability of financial resources. BoP users have a lesser amount of money to spend, and hence, financial attributes are one of the apparent criteria to consider when selecting a technology. The attributes (Fig. 3.1) can be extracted from various literatures on BoP users and markets.

Almost all of the design approaches emphasise the point that affordability must be the primary objective while designing for the BoP (Castillo et al. 2012; Prahalad 2005). Other than the selling price of the product, affordability can also mean the financial model of sales, such as Equated Monthly Instalments (EMIs) and rental services (Castillo et al. 2012; Prahalad 2005). Maintenance costs, along with the replacement costs for the consumable items, also add up to the user's

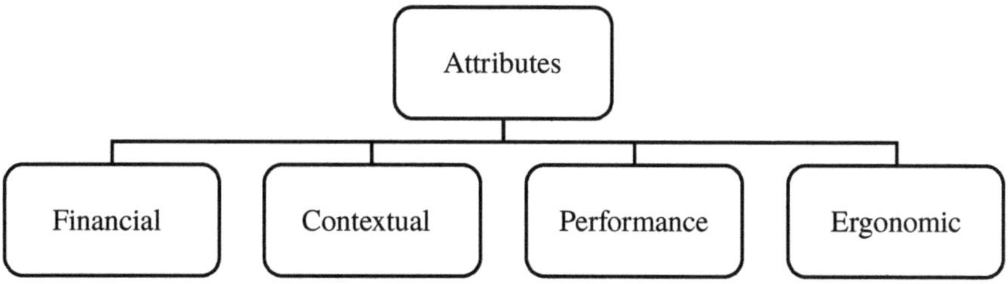

Fig. 3.1 Classification of attributes to be considered

financial burden (Andersen and Kim 2011). In some cases, the user may incur extra expenses if the installation is not easy or if the installation demands skilled labour and expensive tools (Prahalad 2005; Fisher 2007). After-sales support is also an important criterion in technology adoption (Khadilkar 2017; Donaldson 2006). To reduce the financial risks associated with the technology, designers have come up with modular products (Wood et al. 2012). This allows the user to invest in new technology as per their capabilities gradually.

The second factor to be considered while selecting a technology is feasibility. From the BoP user's perspective, feasibility refers to appropriate technology (Castillo et al. 2012). Here, context-specific attributes are used to evaluate the technology. The contextual attributes reflect the requirements that may be unique to a particular user. The third set of attributes pertains to the performance of the technology. These are the standard performance metrics used to evaluate a given technology. Ergonomics is the fourth factor considered to identify the attributes. The importance of ergonomics in the context of BoP has been discussed in various literatures (Fisher 2007; Sijali and Mwago 2011). The ergonomic considerations should be based on the target user population. Failing to do so can have detrimental effects on the long-term usage of the device (Jagtap et al. 2014; van der Bijl-Brouwer and Dorst 2017; Allen et al. 2017).

3.2.3 *Ranking*

TOPSIS-based approaches have been employed in order to rank the technology alternatives. This method is useful when the attributes have different units and cannot be compared directly. The attributes are assigned weights based on the user scenarios or the opinions of decision-makers. The normalised decision matrices are then used to derive the hypothetical ideal solution. The alternatives are ranked based on the Euclidean distance from the ideal solution. The closer the alternative is to the ideal solution, the higher the ranking. The general steps for ranking alternatives using TOPSIS are explained next.

3.2.3.1 Technology Database

A database containing all the alternatives of a particular technology is created. The database should contain the maximum possible information for each alternative. Only the attributes for which data is available for all the alternatives are considered for the ranking.

3.2.3.2 Elimination Criteria

Some of the alternatives from the database can be eliminated based on user requirements and constraints. Users can also define threshold values of certain attributes. Alternatives not fulfilling such criteria or thresholds can be eliminated from the database.

3.2.3.3 Decision Matrix

The information from the database is written in the form of a decision matrix. Each element d_{ij} of the matrix **D** (Eq. 3.4) represents the jth attribute of the ith alternative. Quantitative attributes can be directly written in the matrix, whereas qualitative attributes have to be converted using the Likert scale (Joshi et al. 2015) or represented using fuzzy numbers. The decision matrix **D** is normalised to form the normalised decision matrix **N**. This is done so that the values are comparable to each other. The element n_{ij} of the matrix **N**, is given by

$$n_{ij} = \frac{d_{ij}}{\| d_{ij} \|} \tag{3.8}$$

where

$$\| d_{ij} \| = \left(\sum_{i=1}^{m} \sum_{j=1}^{n} d_{ij}^2 \right)^{1/2} \tag{3.9}$$

3.2.3.4 Scenarios and Weights

Not all attributes contribute equally to the ranking and selection of the technology. The attributes can have varying degrees of importance as per the user's context. In order to capture this, weights are assigned to each of the attributes. Methods such as user interviews, scenarios, and decision-makers can be used to extract the importance of attributes and assign weights. Scenarios represent the actual situation and context of the user. User scenarios contain the background information of the user in terms of socio-economic, demographics, and geographic conditions. The relative importance of attributes can be extracted from the scenarios. In the MADM techniques, attribute weights are extracted from the Relative Importance Matrix **R**. This matrix is obtained by considering two attributes at a time and deciding on their relative importance. The elements of Relative Importance Matrix **R**, denoted as r_{ij} is given by

$$r_{ij} = \frac{\text{importance of the } i\text{th attribute}}{\text{importance of the } j\text{th attribute}} \tag{3.10}$$

If n attributes are considered, the matrix will be of the order $n \times n$. The diagonal elements of the matrix **R** are 1, and the off-diagonal elements will be reciprocals of each other. The relative importance matrix can be written as

$$\mathbf{R} = \begin{bmatrix} 1 & r_{12} & \cdots & r_{1n} \\ r_{21} & 1 & \cdots & r_{2n} \\ \vdots & \vdots & \ddots & \vdots \\ r_{n1} & r_{n2} & \cdots & 1 \end{bmatrix} \tag{3.11}$$

Subsequently, weights of individual attributes can be obtained by using the eigenvector method. If an eigenvalue of **R** is denoted with λ_e and the corresponding eigenvector is **w**, then

$$\mathbf{R}\mathbf{w} = \lambda_e \mathbf{w} \tag{3.12}$$

which implies

$$(\mathbf{R} - \lambda_e \mathbf{1})\mathbf{w} = \mathbf{0} \tag{3.13}$$

where **1** is the $n \times n$ identity matrix and **0** is the n-dimensional vector of zeros. The eigenvector **w** is then obtained by normalising it such that

$$\sum_{i=1}^{n} w_i = 1 \tag{3.14}$$

where w_i is the ith element of an eigenvector **w**. For the largest eigenvalue of λ_e in Eq. 3.12, the elements of the corresponding eigenvector **w**, namely $w_1, w_2, \ldots, w_n$, constitute the weights of the attributes.

It may be noted that the relative importance of the attributes varies for different users based on various socio-economic and technical factors. Hence, multiple scenarios can be considered to rank the devices, wherein each scenario represents a set of attribute weights. The weight vector **w** is then multiplied to the normalised decision matrix **N** to obtain the weighted normalised matrix **V**.

3.2.3.5 Benchmarking and Ranking

The TOPSIS method ranks the options based on *relative closeness to the ideal solution*. The ideal solutions are the benchmarks that can be obtained by the Weighted Normalised Matrix **V**. Benchmarks are the hypothetical alternatives that have the best and the worst attributes. The hypothetical alternative with the best value for each attribute is called the positive benchmark ($\mathbf{v}^+$), and the one with the worst values is called the negative benchmark ($\mathbf{v}^-$).

The positive and negative benchmarks are written by selecting the largest and smallest values from each column corresponding to an attribute. The ranking is then done based on C^*, which is the measure of relative closeness to the ideal solution, as shown in Eqs. 3.5–3.7.

3.3 Case Studies

Here we consider the ranking and selection of three technologies—treadle pump, street vending cart, and *Tulsi mala* bead-making device. Treadle pump and *Tulsi mala* bead-making devices are mainly used in rural areas, whereas the street vending cart is seen mostly in the urban streets. However, all these devices are designed for the BoP. The differences can be seen in the choice of attributes and the relative importance given to them. The case studies presented follow different approaches to assigning weights. For the treadle pump and *Tulsi mala* devices, crisp attributes are considered, and the weights are assigned based on user scenarios. For the vending cart, fuzzy attributes are considered, and decision-makers are involved in the ranking.

3.3.1 Case Study 1: Treadle Pump

To illustrate the ranking of treadle pumps, let us consider five different types. The differences here can be in terms of the type of pumping, materials used, pumping depth, etc. Attributes are identified, and ranking was done based on the TOPSIS method. Scenarios were used to assign weights to various attributes shown in Table 3.3 based on user preferences.

3.3.1.1 Treadle Pump Attributes

As shown in Table 3.3, attributes are identified for the treadle pumps under four different categories, namely Financial, Contextual, Performance, and Ergonomic.

Table 3.3 Attributes identified for ranking treadle pumps

Financial	Contextual	Performance	Ergonomic
1. Price	11. Pumping type	22. Delivery height	28. Step height
2. Purchase model	12. Mechanism type	23. Pumping depth	29. Effort required
3. Maintenance costs	13. Portability	24. Horizontal range	30. Power required
4. Durability	14. Inlet pipe dimension	25. Discharge	
5. After-sales support	15. Outlet pipe dimension	26. Stroke	
6. Modular design	16. Weight	27. Mechanical advantage	
7. Accessories	17. Materials used		
8. Tool required	18. Footprint		
9. Skill set	19. Ease of maintenance		
10. Construction work	20. Cylinder dimensions		
	21. Factor of safety		

Table 3.4 Quantification of ease of maintenance

Condition	Score
Difficult to dismantle (welded)	1
Easy to dismantle but replacement parts not standard	3
Easy to dismantle and replace with standard parts	5

3.3.1.2 Quantification of Attributes

It can be observed that not all of the attributes can be quantified directly. The qualitative attributes can be assigned scores on a 5-point or 10-point scale based on the user experiences (Bhangale et al. 2004). For instance, ease of maintenance refers to the ease with which the user can repair the product or conduct periodic maintenance. In the case of treadle pumps, the pump body can either be welded as a single unit or be an assembly of various parts. If the pump is a welded unit, any damage to any of the parts could mean that the user has to replace the entire unit, whereas if the parts are assembled by threaded joints, the user can easily dismantle the pump and replace the defective part, thereby saving maintenance costs. Table 3.4 illustrates the quantification of ease of maintenance.

However, it should be noted that quantification of all the attributes is not necessary. Attributes such as portability, pumping type, and mechanism type are meant only to convey information about the pump and do not have any mathematical significance. These attributes can be used as elimination criteria while ranking and selecting the treadle pumps.

3.3.1.3 Database

A database of treadle pumps is to be created in order to rank them. A sample database for the RuTAG IITD treadle pump is shown in Table 3.5. Five different

Table 3.5 Attributes identified for RuTAG IITD treadle pump

Sl. no.	Attribute	Unit	Qualitative data	Quantitative value
Financial				
1	Price	Rs.	–	12,000
2	Purchase model	–	Full payment upfront	1
3	Maintenance costs	Rs./month	–	No info
4	Durability	Years	–	10
5	After-sales support	Months	–	12
6	Modular design	-	No	–
7	Accessories	–	No info	No info
8	Tools required	–	Many	1
9	Skill set	–	Skilled labour required	1
10	Construction work	–	Civil construction required	1
Contextual				
11	Pumping Type	–	Suction	-
12	Mechanism type	–	Rocker-lever	-
13	Portability	–	Fixed	-
14	Inlet pipe dimensions	mm	–	40
15	Outlet pipe dimensions	mm	–	NA
16	Weight	kg	–	30
17	Materials used	–	Mild steel	-
18	Footprint	Sq. mm	–	255×10^3
19	Ease of maintenance	–	Easy to dismantle and standard replacement parts	5
20	Cylinder dimensions	mm	–	89
21	Factor of safety	–	-	No Info
Performance				
22	Delivery height	m	–	7
23	Pumping depth	m	–	7
24	Horizontal range	m	–	0
25	Discharge	*lph*	–	3600
26	Stroke	mm	–	100
27	Mechanical advantage	–	-	No info
Ergonomic				
28	Step height	mm	–	240
29	Effort required	N	–	280
30	Power required	W/BMR	–	No info

treadle pumps were selected (see Table 3.6) and ranked based on six different attributes. The weights of the attributes were deduced based on scenarios.

Table 3.6 Database of treadle pumps

Sl. no.	Treadle pump*	Price (₹)	Discharge (lph)	Pumping depth (m)	Delivery height (m)	Horizontal range (m)	Ease of maintenance
1	Ajay Ind.	4500	1700	7	16	200	1
2	RuTAG IITD	12,000	3600	7	7	0	5
3	Ecoflo	5000	5000	7	15	500	3
4	RuKart	5750	1750	7	12	200	3
5	Moneymaker Max	12,750	3600	7	14	200	1

*Brief details about these pumps are given in Appendix A

3.3.1.4 Decision Matrix

For the current database in Table 3.6, the decision matrix is written as

$$\mathbf{D} = \begin{bmatrix} 1/4500 & 1700 & 7 & 16 & 200 & 1 \\ 1/12000 & 3600 & 7 & 7 & 0 & 5 \\ 1/5000 & 5000 & 7 & 15 & 500 & 3 \\ 1/5750 & 1750 & 7 & 12 & 200 & 3 \\ 1/12750 & 3600 & 7 & 14 & 200 & 1 \end{bmatrix} \tag{3.15}$$

It should be noted that for cost attributes such as price and weight, where lower values are preferred, their reciprocals are written in the decision matrix. Here, in column 1 of matrix **D**, the reciprocals of price are written. For example, the price of the first treadle pump is ₹4500. Hence, its reciprocal is written in the first column as 1/4500. Since the number of treadle pumps is 5 and the attributes considered are 6, the decision matrix **D** is of the size 5 × 6.
The normalised decision matrix can be obtained using Eqs. 3.8 and 3.9.

$$\mathbf{N} = \begin{bmatrix} 0.6099 & 0.2254 & 0.4472 & 0.5425 & 0.3288 & 0.1491 \\ 0.2218 & 0.4774 & 0.4472 & 0.2373 & 0 & 0.7454 \\ 0.5545 & 0.6630 & 0.4472 & 0.5085 & 0.8220 & 0.4472 \\ 0.4713 & 0.2321 & 0.4472 & 0.4068 & 0.3288 & 0.4472 \\ 0.2218 & 0.4774 & 0.4472 & 0.4746 & 0.3288 & 0.1491 \end{bmatrix} \tag{3.16}$$

3.3.1.5 Scenarios

In the case of treadle pumps, scenarios are constructed based on various factors of irrigation. In this study, *type of terrain*, *type of irrigation*, and *type of locality* were considered to construct scenarios.

Terrain can be of two types, plain and hilly. The requirements of the farmer in terms of irrigation vary in both these conditions. For instance, higher importance is given to *delivery height* in the case of hilly terrain, whereas in plain lands *discharge rate* and *horizontal range* are given more importance.

Two types of irrigation are considered, namely piped and channel. For piped irrigation, a pipe or a hose is attached to the outlet of the treadle pump. Only a pressure-type treadle pump can be used in such cases. The farmer gives higher importance to *discharge* and *horizontal range*. In channel irrigation, canals or channels are built in the fields. The water drawn from the treadle pump is directly delivered into one of the channels. In this case, *horizontal range* would be given lower priority by the user.

The type of locality is also an important factor to be considered when choosing a treadle pump. It is relatively easier to obtain spare parts and undertake repairs in urban areas. Hence, we consider two types of localities, rural and suburban. Users in rural areas give higher importance to *ease of maintenance* as compared to the users in suburban areas.

From the factors, eight different scenarios can be constructed, as shown in Table 3.7. Saaty's scale (Saaty 1977) can be used to convert the linguistic importance to numeric values. Each scenario then results in a particular Relative Importance Matrix and a particular set of attribute weights.

The relative importance matrix and, subsequently, attribute weights can be calculated using Eqs. 3.10–3.14, as explained in Sect. 3.2.3. As an illustration, we consider one of the scenarios, say Scenario 4 (plain-channel-rural), to understand the ranking methodology.

In scenario 4, since it is a rural setting, high importance is given to *ease of maintenance*. *Delivery height* is not an important attribute in plain lands. Similarly, for channel irrigation, *horizontal range* is also given lower priority. Using Saaty's scale, values of relative importance for the attributes can be written as shown in Eq. 3.17.

$$
\mathbf{R} = \begin{array}{c|cccccc|}
 & \textit{price} & \textit{discharge} & \begin{array}{c}\textit{pumping}\\ \textit{depth}\end{array} & \begin{array}{c}\textit{delivery}\\ \textit{height}\end{array} & \begin{array}{c}\textit{Horizontal}\\ \textit{range}\end{array} & \begin{array}{c}\textit{Easeof}\\ \textit{maintenance}\end{array} \\
\hline
\textit{price} & 1 & 1/3 & 3 & 7 & 7 & 1/3 \\
\textit{discharge} & 3 & 1 & 5 & 7 & 7 & 1 \\
\begin{array}{c}\textit{pumping}\\ \textit{depth}\end{array} & 1/3 & 1/5 & 1 & 3 & 3 & 1/5 \\
\begin{array}{c}\textit{delivery}\\ \textit{height}\end{array} & 1/7 & 1/7 & 1/3 & 1 & 1 & 1/7 \\
\begin{array}{c}\textit{horizontal}\\ \textit{range}\end{array} & 1/7 & 1/7 & 1/3 & 1 & 1 & 1/7 \\
\begin{array}{c}\textit{Ease of}\\ \textit{maintenance}\end{array} & 3 & 1 & 5 & 7 & 7 & 1
\end{array}
\tag{3.17}
$$

The weight vector can then be obtained as

$$
\mathbf{w} = \begin{bmatrix}0.1809 & 0.3359 & 0.0774 & 0.0349 & 0.0349 & 0.3359\end{bmatrix}^T \tag{3.18}
$$

Table 3.7 Scenarios and importance of attributes

	Scenario 1	Scenario 2	Scenario 3	Scenario 4	Scenario 5	Scenario 6	Scenario 7	Scenario 8
	Plain	Plain	Plain	Plain	Hilly	Hilly	Hilly	Hilly
	Piped	Piped	Channel	Channel	Piped	Piped	Channel	Channel
Attributes	Suburban	Rural	Suburban	Rural	Suburban	Rural	Suburban	Rural
Discharge	High	High	Medium	Medium	Medium	Medium	Medium	Medium
Delivery height	Low	Low	Low	Low	High	High	High	High
Horizontal range	High	High	Low	Low	High	High	Low	Low
Ease of maintenance	Low	High	Low	High	Low	High	Low	High

These weights are then multiplied to the normalised decision matrix **N** to obtain the 6×6 Weighted Normalised Matrix **V**, which is given by

$$\mathbf{V} = \begin{bmatrix} 0.1104 & 0.0757 & 0.0346 & 0.0190 & 0.0115 & 0.0501 \\ 0.0401 & 0.1603 & 0.0346 & 0.0083 & 0 & 0.2504 \\ 0.1003 & 0.2227 & 0.0346 & 0.0178 & 0.0287 & 0.1502 \\ 0.0853 & 0.0779 & 0.0346 & 0.0142 & 0.0115 & 0.1502 \\ 0.0401 & 0.1603 & 0.0346 & 0.0166 & 0.0115 & 0.0501 \end{bmatrix} \tag{3.19}$$

3.3.1.6 Benchmarking and Ranking

The positive and negative benchmarks for this illustration can be written as

$$\mathbf{v}^{+} = \begin{bmatrix} 0.1104 & 0.2227 & 0.0346 & 0.0190 & 0.0287 & 0.2504 \end{bmatrix} \tag{3.20}$$

$$\mathbf{v}^{-} = \begin{bmatrix} 0.0401 & 0.0757 & 0.0346 & 0.0083 & 0 & 0.0501 \end{bmatrix} \tag{3.21}$$

Now, C^* scores are obtained as

$$\mathbf{C}^* = \begin{bmatrix} 0.2242 & 0.6876 & 0.6539 & 0.3824 & 0.2789 \end{bmatrix}^T \tag{3.22}$$

3.3.1.7 Results and Discussions

Five different treadle pumps were compared and ranked for eight scenarios. Weights of the attributes were calculated from the corresponding Relative Importance Matrix using the eigenvalue method. The elimination criteria were applied in scenarios with piped irrigation, and the suction pumps were eliminated from the database. For instance, in scenarios 1, 2, 5, and 6, the RuTAG IITD treadle pump is not considered for the ranking. The rankings of the treadle pumps across eight scenarios can be seen in Table 3.8. The variations of the rankings of treadle pumps across scenarios can be visually represented, as shown in Fig. 3.2. The C^* scores indicate the distance of an alternative from the ideal solution. It can be observed that in some scenarios, the scores of the alternatives are very close to each other. The C^* scores and rankings are affected by the attribute weights. In order to study the same, a sensitivity analysis was done.

To understand the effects of weights on treadle pump rankings, a weight disturbance is introduced in each scenario, and the change in ranking is observed. If an attribute is more sensitive, the ranking of treadle pumps changes even with a small disturbance. Here, pairs of alternatives as described in Triantaphyllou and Sánchez (1997) were considered, and minimum change in weight was considered as the objective function for our optimisation problem.

Table 3.8 Treadle pump ranking for each scenario

	Scenario 1		Scenario 2		Scenario 3		Scenario 4		Scenario 5		Scenario 6		Scenario 7		Scenario 8	
	Plain		Plain		Plain		Plain		Hilly		Hilly		Hilly		Hilly	
	Piped		Piped		Channel		Channel		Piped		Piped		Channel		Channel	
Treadle pump	Suburban		Rural		Suburban		Rural		Suburban		Rural		Suburban		Rural	
Ajay Ind.	C*	Rank	C*	Rank	C*	Rank	C*	Rank	C*	Rank	C*	Rank	C*	Rank	C*	Rank
	0.2318	3	0.1931	3	0.3056	4	0.2242	5	0.2677	2	0.2229	3	0.4231	4	0.3204	5
RuTAG IITD	NA	NA	NA	NA	0.5226	2	0.6876	1	NA	NA	NA	NA	0.4352	3	0.6160	2
Ecoflo	0.9597	1	0.9674	1	0.8274	1	0.6539	2	0.9333	1	0.9462	1	0.8468	1	0.6724	1
RuKart	0.2499	2	0.4610	2	0.2730	5	0.3824	3	0.2286	3	0.4566	2	0.3222	5	0.3976	3
Moneymaker Max	0.1802	4	0.1478	4	0.4261	3	0.2789	4	0.1850	4	0.1511	4	0.5220	2	0.3479	4

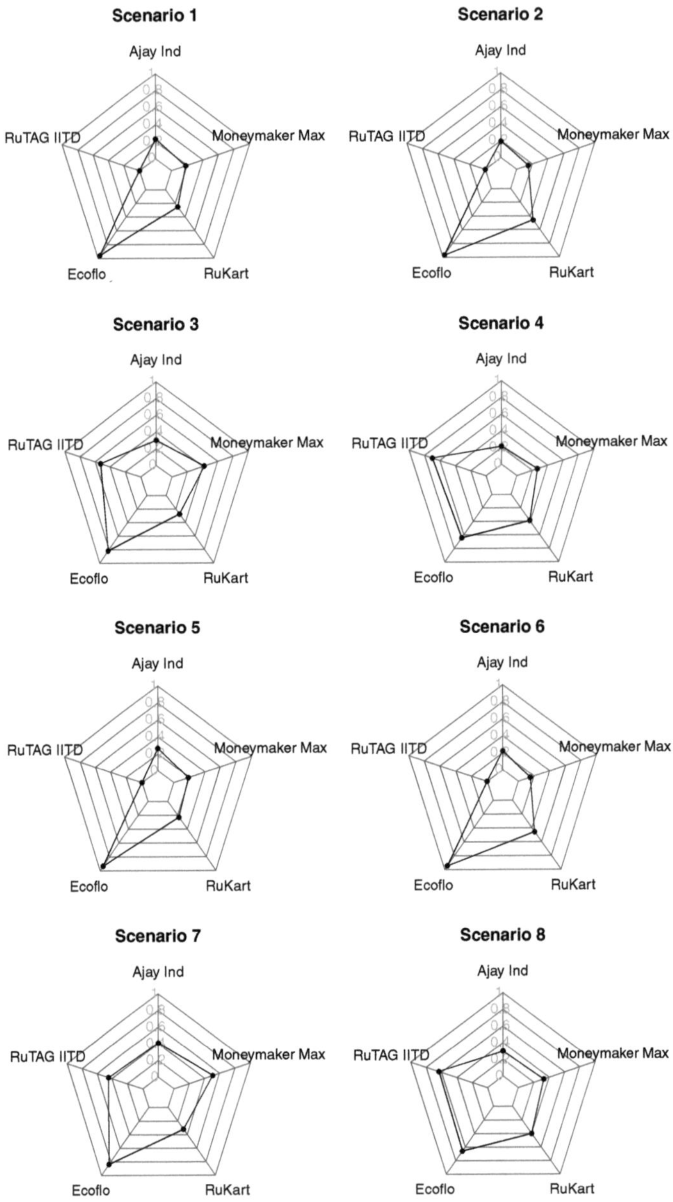

Fig. 3.2 Comparison of treadle pumps

If the weight of nth attribute w_n is changed to w_n^*, the corresponding changes in the weights of other attributes can be calculated as given in Li et al. (2013).

$$w_i^* = \frac{w_i}{1 + (\gamma - 1)w_n} \tag{3.23}$$

where

$$\gamma = \frac{w_n^*}{w_n} \tag{3.24}$$

The optimisation function to obtain the changed weight of the nth attribute can be formulated as follows:

Minimise

$$f(x) = w_n^* - w_n \tag{3.25}$$

subject to

$$C^*(v) - C^*(u) = 0 \tag{3.26}$$

where u, v are the alternative pairs.

The results of the sensitivity analysis are shown in Table 3.9. Bold-faced values indicate the most sensitive attribute in each scenario. For instance, in Scenario 1, the ranking of treadle pumps changes if the weight assigned to the attribute *ease of maintenance* is reduced by 12.42% or more. On the other hand, NC indicates that the attributes are not critical in that particular scenario, which means that any change in the weight of those attributes does not affect the ranking of the treadle pumps. The attribute *pumping depth* is not critical for any of the scenarios. This is because the values for *pumping depth* are the same for all the alternatives of treadle pumps, and any change in the weights will not affect the ranking. Similarly, the attribute *horizontal range* is not critical in scenarios where channel irrigation is used, as the weight assigned to it is very low.

3.3.2 *Case Study 2: Tulsi Mala Bead-Making Device*

As seen in the previous chapter, the specifications and design elements of the bead-making devices are different, and hence, a particular device might be appropriate only for a specific use case. Here, five different *Tulsi mala* bead-making devices are ranked and compared for various user scenarios.

Table 3.9 Sensitivity analysis of attributes

Attributes	Scenario 1	Scenario 2	Scenario 3	Scenario 4	Scenario 5	Scenario 6	Scenario 7	Scenario 8
Price	+14.68	**−24.31**	−26.47	+20.95	NC	+308.48	+10.47	+33.16
Discharge	+26.26	+29.17	**−21.93**	+13.34	+32.34	+63.76	**−2.84**	**−12.17**
Pumping depth	NC	NC	NC	NC	NC	NC	NC	NC
Delivery height	+397.9	+1850.9	+500	+349.28	−32.23	+133.19	+2.78	−28.27
Horizontal range	NC	NC	+217.17	+97.13	NC	NC	+87.65	+843.46
Ease of maintenance	**−12.42**	−65.14	+28.36	**−7.08**	**+26.14**	**−54.98**	−9.6	+12.89

3.3.2.1 Attributes

The *Tulsi mala* bead-making devices are primarily used in the villages, and the artisans depend on them to earn their livelihood. Various socio-economic factors are to be considered while identifying the relevant attributes on which the alternatives will be compared. The identified attributes are shown in Table 3.10.

3.3.2.2 Quantifying Attributes

Tables 3.11, 3.12, and 3.13 illustrate the quantification of some of the qualitative attributes.

It can be noted that attributes like *time to learn*, *skill set*, and *effort required* should have lower values. Hence, higher scores are given to lower values. Such attributes are called cost attributes, whereas *comfort* is a benefit attribute, which should have a higher value. There can also be non-monotonic attributes where the preferred values lie in the middle. *Operating speed* is one such attribute. It can be converted to a 5-point scale, as shown in Table 3.14.

Table 3.10 Attributes for *Tulsi mala* bead-making devices

Financial	Contextual	Performance	Ergonomic
1. Price	7. Skill set	16. Bead finish	20. Comfort
2. Purchase model	8. Portability	17. Bead size	21. Noise and vibration
3. Maintenance costs	9. Weight	18. Material wastage	22. Effort required
4. Accessories	10. Footprint	19. Reliability	23. Ease of handling and assembly
5. Earnings	11. Ease of maintenance		24. Safety
6. Power consumption	12. Operating speed		
	13. Materials used		
	14. Time to learn		
	15. Energy source		

Table 3.11 Quantification of *time to learn*

Time to learn	Score (5-point scale)
Long	1
Moderate	3
Quick	5

Table 3.12 Quantification of *skill set* and *effort required*

Skill set/ Effort required	Score (5-point scale)
High	1
Medium	3
Low	5

Table 3.13 Quantification of *comfort*

Comfort	Score (5-point scale)
Very low	1
Low	2
Medium	3
High	4
Very high	5

Table 3.14 Quantification of *operating speed*

Operating speed (RPM)	Score (5-point scale)
<1000 or >8000	1
1000–1500 or 2500–8000	3
1500–2500	5

3.3.2.3 Decision Matrix

The database considered for this study is shown in Table 3.15.

The decision matrix can be written as follows:

$$\mathbf{D} = \begin{bmatrix} 1 & 5 & 1 & 1 & 1 & 2 \\ 1.5 & 3 & 3 & 3 & 5 & 2 \\ 0.125 & 3 & 3 & 3 & 5 & 3 \\ 0.2 & 3 & 1 & 1 & 5 & 1 \\ 0.067 & 5 & 5 & 5 & 5 & 5 \end{bmatrix} \tag{3.27}$$

It is to be noted that the attribute *bead size* is a filtering attribute and does not have any mathematical significance in the decision-making process. Hence, it does not appear in the decision matrix. The normalised decision matrix N is written as

$$\mathbf{N} = \begin{bmatrix} 0.6175 & 0.569 & 0.149 & 0.149 & 0.099 & 0.305 \\ 0.7719 & 0.342 & 0.447 & 0.447 & 0.497 & 0.305 \\ 0.0772 & 0.342 & 0.447 & 0.447 & 0.497 & 0.457 \\ 0.1235 & 0.342 & 0.149 & 0.149 & 0.497 & 0.152 \\ 0.0414 & 0.569 & 0.745 & 0.745 & 0.497 & 0.762 \end{bmatrix} \tag{3.28}$$

3.3.2.4 Scenarios

For the case of *Tulsi mala* bead-making device, three factors were considered to create scenarios. They are the type of beads, the type of artisan, and the skill level of the artisan.

Type of Beads

The sizes of the beads vary, and also, the beads are used in different settings and applications. The range of bead sizes that a device can produce also varies. Artisans

Table 3.15 Database of *Tulsi mala* bead-making device

Device	Price (₹)	Operating speed	Bead size range (mm)	Time to learn	Skill set	Effort required	Comfort
Hand-operated traditional	1000	5	2–15	1	1	1	2
Handheld DC motor	800	3	2–3	3	3	5	2
Sliding DC motor	8000	3	4–10	3	3	5	3
Improvised AC motor	5000	3	8–15	1	1	5	1
Pedal-operated AC motor	15,000	5	6–15	5	5	5	5

produce four types of beads using these devices. *Kanthi mala* beads are the smallest beads used to make necklaces or *kanthi malas*. Their sizes vary from 2 *mm* to 5 *mm* in diameter. The next size of the bead is called *matar dana* or pea-sized bead. They are 5 *mm* to 8 *mm* in diameter. *Japa mala* beads are of size 8 *mm* to 10 *mm* in diameter. Larger beads of sizes ranging from 10 *mm* to 15 *mm* in diameter are also produced. *Malas* made of these beads are often used as offerings to the deity.

Type of Artisan

It can be found that in a typical household, men, women, and children are involved in the bead-making process. Since the physical strength varies across these categories, not all artisans can work comfortably on all the devices. Some devices demand very high physical strength, whereas some require minimal effort.

Skill of the Artisan

Similar to physical strength, the skill level of the artisans varies from one another. Artisans can be classified into beginner, intermediate, and expert categories based on their skill levels. The time to learn to operate and get accustomed to a device varies for each alternative. For a beginner, it is important that the learning curve of a device is not very steep.

The factors are summarised in Table 3.16. Using different combinations of these factors, a total of 36 scenarios can be created. Each scenario results in a particular Relative Importance Matrix and a particular set of attribute weights. For instance, a woman artisan with an expert skill level would give higher importance to effort and comfort and lower importance to time to learn and skillset. In contrast, a man with the skill level of a beginner would give higher importance to time to learn and skill set.

3.3.2.5 Results and Discussions

Five *Tulsi mala* bead-making devices are ranked for various scenarios. It can be observed that the rankings vary for each of the scenarios. Results for four different scenarios are shown in Figs. 3.3, 3.4, 3.5, and 3.6.

Table 3.16 Factors to create scenarios

Type of beads	Type of artisan	Skill level
Kanthi mala	Men	Beginner
Matar dana	Women	Intermediate
Japa mala	Children	Expert
Large beads		

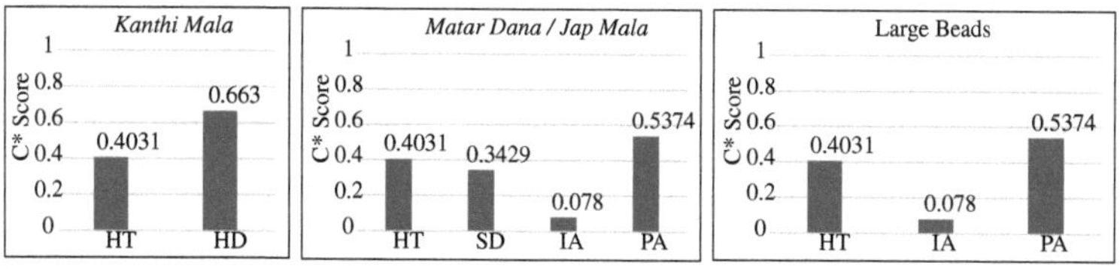

HT:Hand-operated Traditional ; HD:Handheld DC motor ; SD:Sliding DC motor ; IA:Improvised AC motor; PA:Pedal-operated AC motor

Fig. 3.3 Device scores for scenario 1

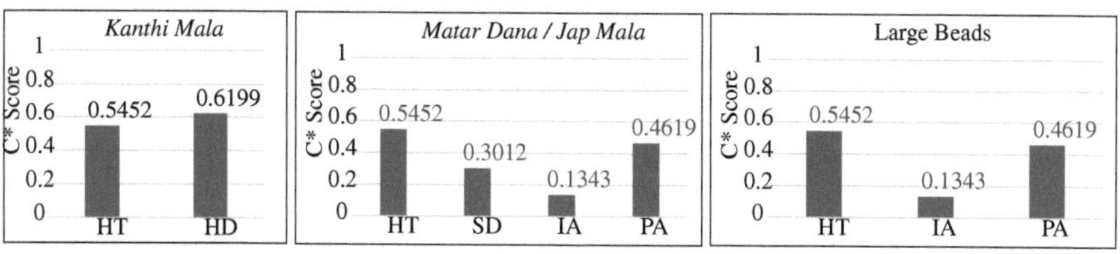

Fig. 3.4 Device scores for scenario 2

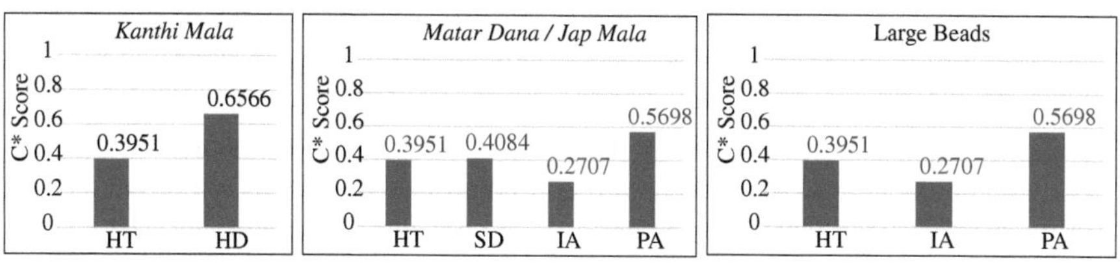

Fig. 3.5 Device scores for scenario 3

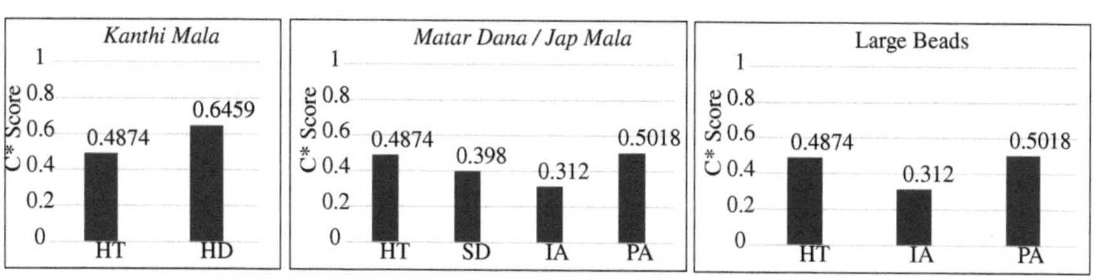

Fig. 3.6 Device scores for scenario 4

Scenario 1: Men with Beginner Skill Level

High importance: *price*, *time to learn*, and *skill set*; medium importance: *operating speed*; low importance: *effort* and *comfort*

Scenario 2: Men with Expert Skill Level

High importance: *price* and *comfort*; medium importance: *operating speed* and *effort*; low importance: *time to learn* and *skill set*

Scenario 3: Women with Beginner Skill Level

High importance: *price*, *time to learn*, *skill set*, *effort*, and *operating speed*; medium importance: *comfort*

Scenario 4: Women with Expert Skill Level

High importance: *price*, *comfort*, and *effort*; medium importance: *operating speed*; low importance: *time to learn* and *skill set*

It can be noted that not all bead types can be produced using all the devices. Hence, the results are segregated based on the type of beads produced. *Kanthi mala* beads can only be produced using traditional hand-operated and handheld DC motor devices. The results show that the handheld DC motor device is preferable in all the scenarios. It can also be observed in the field that since the traditional hand-operated device demands high skill and effort, it is predominantly used by experienced men. For the medium- and larger-sized beads, i.e. *matar dana*, *jap mala*, and large beads, the pedal-operated AC-powered device is preferable in the case of beginners, and the traditional hand-operated device scores well in the case of expert men. However, for women, it can be seen that the pedal-operated AC-powered device fares well even for experts. It was also observed in the field that women do not prefer the traditional hand-operated device and the improvised AC motor device due to the limitation of physical strength.

3.3.3 Case Study 3: Street Vending Cart

In this study, we look at how scenario-based selection is applied to selecting appropriate fruit and vegetable vending carts. Since multiple stakeholders are usually involved in the selection process, an additional dimension of group decision-making is considered here.

3.3.3.1 Attributes

It is worth noting that many individuals are drawn to the profession of street vending because it requires minimal skill and financial investment (Naik 2013; Kotwal et al. 2019). An average street vendor would need to spend almost a month's earnings

Table 3.17 Attributes identified for ranking vending carts (Bhat et al. 2021)

Financial	Contextual	Performance	Ergonomic
1. Price	6. Cart type	14. Loading capacity	17. Handle height
2. Purchase model	7. Display area	15. Manoeuvrability	18. Pushing force
3. Maintenance costs	8. Weight	16. Shelf life increase	
4. Durability	9. Material used		
5. Modularity	10. Ease of maintenance		
	11. Visual appeal		
	12. Storage space		
	13. Skill set		

Table 3.18 Objective attributes of the vending carts

Vending cart	Price (₹)	Loading capacity (kg)	Display area (m^2)
Four-Wheel Push Cart	10,000	150	1.4
Tricycle Pull Cart	15,000	200	1.1
Rediwala Push Cart	30,000	150	0.84
GoFlow Push Cart	35,000	250	1.1
Motorised e-cart	1,50,000	310	1.86

(approximately ₹10,000 or US$130–US$150) to purchase a basic cart (McKay et al. 2016). It is thus critical that an appropriate selection strategy is employed while choosing the right vending cart. As seen in the previous sections, the attributes of vending carts are identified in four categories, namely Financial, Contextual, Performance, and Ergonomic, as shown in Table 3.17.

3.3.3.2 Ranking and Decision-Making

The vending cart alternatives are then ranked based on their attributes. As an illustration, we choose the following five attributes to rank the carts: *price*, *loading capacity*, *display area*, *manoeuvrability*, and *visual appeal*. It can be observed that out of the five chosen attributes, the first three are objective attributes, i.e. they have a definite value. The objective attributes of the vending carts are shown in Table 3.18.

The last two attributes *manoeuvrability* and *visual appeal* are subjective and cannot be quantified directly. For subjective attributes, Triangular Fuzzy Numbers (TFNs) are used to represent the opinions of the decision-makers.

TFNs are a useful modelling tool to address the intrinsic ambiguity in subjective evaluations. Since TFNs define linguistic terms with three parameters, a lower limit, most likely value, and an upper limit, they are especially well suited for converting the evaluations into quantifiable representations. For attributes such as *manoeuvrability* and *visual appeal*, TFNs can capture the imprecision and variability in human judgement. The linguistic terms to rate the subjective attributes are given in Table 3.19.

Table 3.19 Linguistic terms for rating subjective attributes

Linguistic terms		
Manoeuvrability	Visual appeal	TFN
Very Easy (VE)	Very Attractive (VA)	(0.8, 1, 1)
Easy (E)	Attractive (A)	(0.6, 0.8, 1)
Moderate (M)	Neutral (N)	(0.3, 0.5, 0.7)
Difficult (D)	Unattractive (U)	(0, 0.2, 0.4)
Very Difficult (VD)	Very Unattractive (VU)	(0, 0, 0.2)

Table 3.20 TFN for rating the importance

Linguistic terms	TFN
Very Important (VI)	(0.7, 1, 1)
Important (I)	(0.5, 0.7, 1)
Medium (M)	(0.2, 0.5, 0.8)
Unimportant (U)	(0, 0.3, 0.5)
Very Unimportant (VU)	(0, 0, 0.3)

A. Decision-Makers

We use three decision-makers to illustrate the group decision-making process. Decision-makers are responsible for evaluating the options based on subjective factors and determining the importance of each attribute. It is also to be noted that, usually, in a group decision-making setup, the importance of each decision-maker also varies. The importance levels of attributes and decision-makers can be represented using the linguistic terms as shown in Table 3.20. The rating scale and the membership function used for the fuzzy numbers are as shown in Liang and Wang (1993).

Corresponding weights are then assigned to the decision-makers as per the importance level. For instance, let the importance levels of three decision-makers be set as Very Important, Important, and Medium, respectively. The decision-makers are then asked to rate the subjective attributes.

B. Decision Matrix

The decision matrix is constructed by using both objective and subjective attributes. For the subjective attributes, the average ratings can be calculated using Eq. 3.29.

$$D_{ij} = \frac{\sum_{i=1}^{k} (P_i \otimes R_{ij})}{\sum_{i=1}^{k} P_i} \tag{3.29}$$

Here:

k: number of decision-makers
P_i: the TFN corresponding to the importance of the ith decision-maker
R_{ij}: the TFN corresponding to the importance of the jth attribute given by the ith decision-maker

Table 3.21 Attribute weights

Attribute	DM1	DM2	DM3	Average weight
Price	I	VI	VI	(0.6, 0.86, 1)
Loading capacity	I	VI	I	(0.57, 0.8, 1)
Display area	VI	VI	I	(0.67, 0.93, 1)
Manoeuvrability	M	U	M	(0.13, 0.44, 0.69)
Visual appeal	M	M	I	(0.24, 0.54, 0.86)

DM decision-makers

The symbol $\otimes$ represents the multiplication of fuzzy numbers. For two TFNs, $U = (a_1, b_1, c_1)$ and $V = (a_2, b_2, c_2)$, $U \otimes V$ is given by $(a_1.a_2, b_1.b_2, c_1.c_2)$.

The normalised decision matrix **N** can be written by normalising the values of objective attributes in each column.

$$\mathbf{N} = \begin{bmatrix} 1 & 0.484 & 0.75 & (0, 0.06, 0.27) & (0.4, 0.56, 0.8) \\ 0.667 & 0.645 & 0.6 & (0.41, 0.56, 0.81) & (0.3, 0.5, 0.7) \\ 0.333 & 0.484 & 0.45 & (0.3, 0.5, 0.7) & (0.7, 0.91, 1) \\ 0.286 & 0.806 & 0.6 & (0.15, 0.34, 0.51) & (0.77, 0.95, 1) \\ 0.066 & 1 & 1 & (0.73, 0.94, 1) & (0.73, 0.94, 1) \end{bmatrix} \tag{3.30}$$

Weights for each attribute are calculated based on the weighted average of importance determined by the decision-makers. An example is shown in Table 3.21.

Weights of the individual attributes are multiplied to the corresponding columns of **N** to obtain the Weighted Decision Matrix **V**. The operator $\otimes$ is used to obtain weighted TFN for subjective attributes.

The Weighted Decision Matrix **V** can be written as

$$\mathbf{V} = \begin{bmatrix} (0.6, 0.86, 1) & (0.28, 0.39, 0.48) & (0.5, 0.7, 0.75) & (0, 0.03, 0.19) & (0.1, 0.3, 0.7) \\ (0.4, 0.57, 0.67) & (0.37, 0.52, 0.64) & (0.4, 0.56, 0.6) & (0.05, 0.25, 0.56) & (0.07, 0.27, 0.6) \\ (0.2, 0.29, 0.33) & (0.28, 0.39, 0.48) & (0.3, 0.42, 0.45) & (0.04, 0.22, 0.48) & (0.17, 0.22, 0.86) \\ (0.17, 0.25, 0.29) & (0.46, 0.64, 0.81) & (0.4, 0.56, 0.6) & (0.02, 0.15, 0.35) & (0.18, 0.51, 0.86) \\ (0.04, 0.06, 0.06) & (0.57, 0.8, 1) & (0.67, 0.93, 1) & (0.09, 0.41, 1) & (0.18, 0.51, 0.86) \end{bmatrix} \tag{3.31}$$

C. Benchmarking and Ranking

The highest and the lowest TFNs from each column of the Decision Matrix form the positive ideal solution (v^+) and the negative ideal solution (v^-).

$$v^+ = \{(0.6, 0.86, 1), (0.57, 0.8, 1), (0.67, 0.93, 1), (0.09, 0.41, 1), (0.18, 0.51, 0.86)\} \tag{3.32}$$

$$v^- = \{(0.04, 0.06, 0.06), (0.28, 0.39, 0.48), (0.3, 0.42, 0.45), (0, 0.03, 0.19), (0.1, 0.3, 0.7)\} \tag{3.33}$$

The separation measures are calculated by finding the Euclidean distance of each alternative from the ideal solutions. The separation measures S_i^* and S_i^- can be obtained as given in Boran et al. (2009):

$$S_i^* = \sqrt{\frac{1}{3n}\sum_{j=1}^{n}[(u_{ij} - u_j^*)^2 + (v_{ij} - v_j^*)^2 + (w_{ij} - w_j^*)^2]} \quad (3.34)$$

$$S_i^- = \sqrt{\frac{1}{3n}\sum_{j=1}^{n}[(u_{ij} - u_j^-)^2 + (v_{ij} - v_j^-)^2 + (w_{ij} - w_j^-)^2]} \quad (3.35)$$

where $(u_i j, v_i j, w_i j)$ is the TFN corresponding to the jth attribute of the ith alternative, and (u_j^*, v_j^*, w_j^*) is the TFN corresponding to the ith attribute of the ideal solution. The alternatives were ranked by the relative closeness to the ideal solution C^*, which can be computed using

$$C^* = \frac{S_i^-}{S_i^* + S_i^-} \quad (3.36)$$

Separation measures and the relative closeness of the alternatives are given in Table 3.22.

3.3.3.3 Results and Discussions

In this case study, the multi-attribute decision-making framework was used to select the most appropriate vending cart for BoP users. An illustration of how subjective attributes can be incorporated in the framework is provided. Five different vending carts were compared across five attributes. Out of which, three attributes (*price, loading capacity, and display area*) were objective, and two (manoeuvrability and visual appeal) were subjective.

The handling of subjective attributes is demonstrated through the use of a group decision-making approach involving three decision-makers with importance levels of Very Important, Important, and Medium. The subjective attributes were assessed by the decision-makers using linguistic terms. TFNs were then used to quantify the

Table 3.22 Separation measures and relative closeness

Vending cart	S^*	S^-	C^*
Four-Wheel Push Cart	0.322	0.369	0.534
Tricycle Pull Cart	0.284	0.266	0.484
Rediwala Push Cart	0.412	0.146	0.262
GoFlow Push Cart	0.365	0.178	0.328
Motorised e-Cart	0.349	0.375	0.518

ratings. The final scores indicated by C^* in Table 3.22 yield the ranking of the carts. While the final rankings are illustrative, the study highlights how this framework of scenario-based ranking can be effectively used in real-world BoP contexts.

3.4 Technology Selection Toolbox

A C#-based tool was developed to assist the users and practitioners in practically implementing the multi-attribute decision-making framework. This toolbox provides a structured environment for evaluating and ranking technology alternatives, enabling context-specific decision-making. The workflow of the tool is as follows: In the start screen Fig. 3.7, the user selects one of the technologies displayed. In this case, three BoP products, treadle pump, Tulsi mala bead-making device, and street vending cart, are displayed. Once the technology to be ranked is chosen, the next screen displays the built-in scenarios (Fig. 3.8). For advanced applications, the user also has the option to create custom scenarios (Fig. 3.9). For instance, in the case of the treadle pump, the user can choose filtering attributes such as *pumping type* and *portability*. The attributes on which the pumps are to be ranked can be chosen on the same screen. Further, the relative importance of the attributes can be entered in the table provided on the screen. It is to be noted that only upper triangular values need to be entered. The rest of the values are generated automatically when the user clicks *set values*. The rankings can be obtained by pressing the *Get Ranking* button as shown in Fig. 3.10. The ranking page also displays the C* scores along with the individual ranks. Due to the modular nature of the tool, it can be used in wider cases of technologies and products built for BoP.

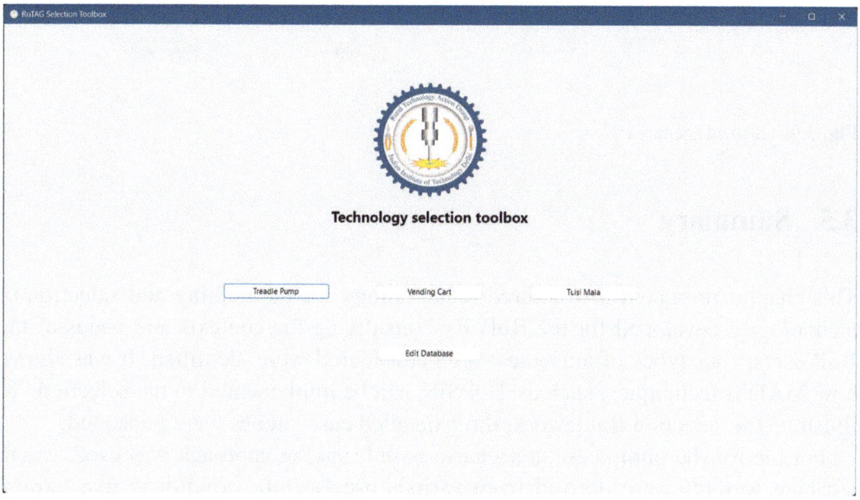

Fig. 3.7 Technology selection page

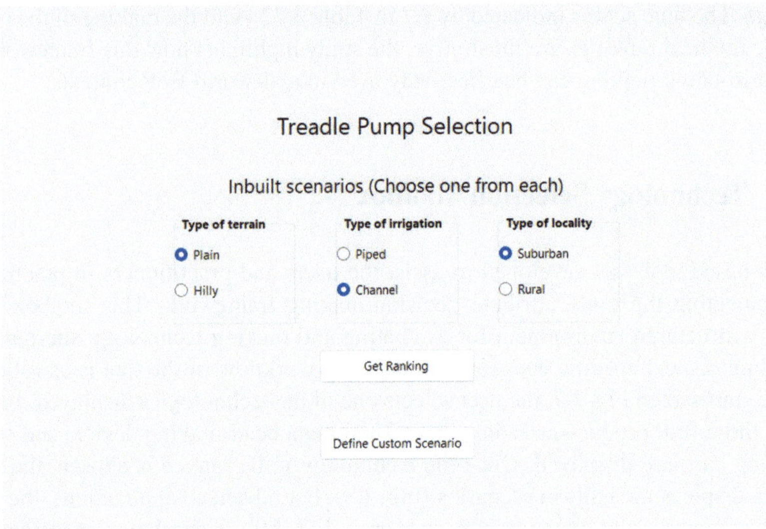

Fig. 3.8 Built-in scenario selection

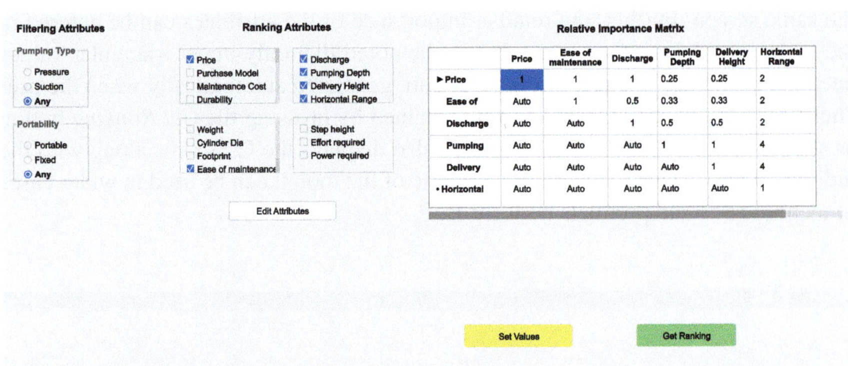

Fig. 3.9 Custom scenario

3.5 Summary

This chapter presented a structured methodology for the ranking and selection of technologies developed for the BoP. By considering the contexts and needs of the BoP users, four types of attributes to be considered were identified. It was shown how MADM techniques, such as TOPSIS, can be implemented in the selection. To illustrate the selection framework, three detailed case studies were presented.

For the treadle pump case, a scenario-based ranking approach was used, where attribute weights were derived from various user-specific conditions like terrain, irrigation method, and locality. A sensitivity study was also conducted to understand the critical attributes in each scenario.

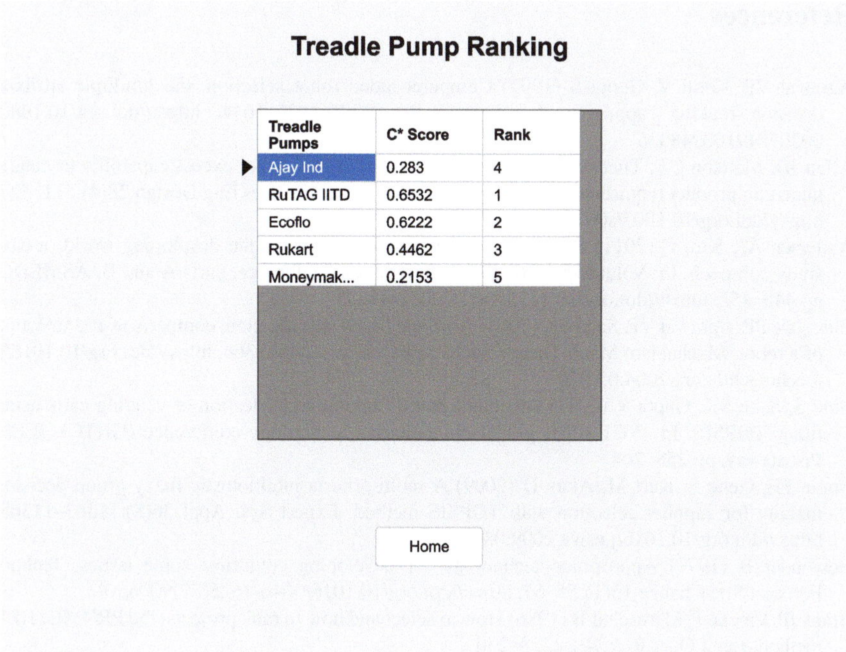

Fig. 3.10 Treadle pump ranking

For the *Tulsi mala* bead-making devices, the methodology was extended to account for non-monotonic attributes, such as optimal operating speed. Here, the derived rankings were cross-validated with field observations. Device preferences predicted by the model, such as lower desirability for hand-operated models by women, matched actual usage trends.

In the case of vending carts, a group decision-making scenario was considered. The subjective attributes were represented using triangular fuzzy numbers. The weights were derived from the opinions of the decision-makers.

To support broader application, a C#-based technology selection toolbox was developed. The tool allows users to rank technology alternatives using the TOPSIS methodology. Users can either use built-in scenarios or create custom scenarios to match the exact field requirements.

While the data used in these case studies were obtained from various sources such as field interviews, market research, and product brochures, it is assumed that the responses given by the users are true and accurate. However, in order to validate the results and test the effectiveness of the rankings, one would need to implement the same in the field and analyse the results. In some cases, such as the *Tulsi mala* bead-making device, the rankings obtained from the proposed methodology were reflected in the field in terms of device usage.

This chapter thus offers both methodological foundation and practical tools for selecting appropriate technologies in BoP settings.

References

Agrawal VP, Kohli V, Gupta S (1991) Computer aided robot selection: the 'multiple attribute decision making' approach. Int J Prod Res 29(8):1629–1644. https://doi.org/10.1080/00207549108948036

Allen JD, Mattson CA, Thacker KS, Ferguson SM (2017) Design for excess capability to handle uncertain product requirements in a developing world setting. Res Eng Design 28(4):511–527. https://doi.org/10.1007/s00163-017-0253-8

Andersen AJ, Kim C (2011) Principles of mechanical design for the developing world: a case study approach. In: Volume 5: 37th design automation conference, parts A and B, ASMEDC, pp 445–452. https://doi.org/10.1115/DETC2011-48245

Bhangale PP, Agrawal VP, Saha SK (2004) Attribute based specification, comparison and selection of a robot. Mechanism Mach Theory 39(12 SPEC. ISS.):1345–1366. https://doi.org/10.1016/j.mechmachtheory.2004.05.020

Bhat S, Saha SK, Gupta V (2021) Attributes based ranking and selection of vending carts using fuzzy TOPSIS. In: 2021 IEEE global humanitarian technology conference (GHTC). IEEE, Piscataway, pp 258–264

Boran FE, Genç S, Kurt M, Akay D (2009) A multi-criteria intuitionistic fuzzy group decision making for supplier selection with TOPSIS method. Expert Syst Appl 36(8):11363–11368. https://doi.org/10.1016/j.eswa.2009.03.039

Bowonder B (1979) Appropriate technology for developing countries: some issues. Technol Forecast Soc Change 15(1):55–67. https://doi.org/10.1016/0040-1625(79)90065-9

Brans JP, Vincke P, Mareschal B (1986) How to select and how to rank projects: the PROMETHEE method. Eur J Oper Res 24(2):228–238

Bødker S (2000) Scenarios in user-centred design—setting the stage for reflection and action. Interact Comput 13(1):61–75. https://doi.org/10.1016/S0953-5438(00)00024-2

Castillo LG, Diehl JC, Brezet JC (2012) Design considerations for base of the pyramid (BoP) projects. In: Proceedings of the Cumulus Helsinki 2012 conference, Helsinki, pp 1–15

Cooper A, Saffo P (1999) The inmates are running the asylum. Macmillan Publishing, USA

Date A (1984) Understanding appropriate technology. In: Appropriate technology in third world development, pp 163–183

Donaldson KM (2006) Product design in less industrialized economies: constraints and opportunities in Kenya. Res Eng Design 17(3):135–155. https://doi.org/10.1007/s00163-006-0017-3

Fisher M (2007) Design to kickstart incomes. Design for the other 90

Floyd IR, Cameron Jones M, Twidale MB (2008) Resolving incommensurable debates: a preliminary identification of persona kinds, attributes, and characteristics. Artifact 2(1):12–26. https://doi.org/10.1080/17493460802276836

Gandhi MK (1929) Urban v. Rural. Young India, pp 244–245

Grudin J, Pruitt J (2002) Personas, participatory design and product development: an infrastructure for engagement. In: Proceedings of the participatory design conference, Palo Alto

Guesalaga R, Marshall P (2008) Purchasing power at the bottom of the pyramid: differences across geographic regions and income tiers. J Consum Market 25(7):413–418. https://doi.org/10.1108/07363760810915626

Huber GP (1974) Multi-attribute utility models: a review of field and field-like studies. Manag Sci 20(10):1393–1402

Hwang CL, Yoon K (1981) Multiple attribute decision making. Lecture notes in economics and mathematical systems, vol 186. Springer, Berlin/Heidelberg. https://doi.org/10.1007/978-3-642-48318-9

Jagtap S, Larsson A, Hiort V, Olander E, Warell A, Khadilkar P (2014) How design process for the base of the pyramid differs from that for the top of the pyramid. Design Stud 35(5):527–558. https://doi.org/10.1016/j.destud.2014.02.007

Joshi A, Kale S, Chandel S, Pal DK (2015) Likert scale: explored and explained. Br J Appl Sci Technol 7(4):396

Kalbar PP, Karmakar S, Asolekar SR (2012) Selection of an appropriate wastewater treatment technology: a scenario-based multiple-attribute decision-making approach. J Environ Manag 113:158–169. https://doi.org/10.1016/j.jenvman.2012.08.025
Kandachar P, Halme M (2005) Introduction: an exploratory journey towards the research and practice of the base of the pyramid. Greener Manag Int 2005(51):3–17. https://doi.org/10.9774/GLEAF.3062.2005.au.00002
Khadilkar P (2017) Formulating the design scope for the base of the (economic) pyramid. Design Issues 33(2):4–17. https://doi.org/10.1162/DESI_a_00436
Klein J (2019) Calculating the AHP priority vector. Tech. rep., Carnegie-Mellon University, Pittsburgh
Kotwal V, Satya S, Naik S, Dahiya A, Kumar J (2019) Street food cart design: a critical component of food safety. In: Rural technology development and delivery. Springer, Berlin, pp 263–278
Li P, Qian H, Wu J, Chen J (2013) Sensitivity analysis of TOPSIS method in water quality assessment: I. Sensitivity to the parameter weights. Environ Monitor Assessm 185(3):2453–2461. https://doi.org/10.1007/s10661-012-2723-9
Liang GS, Wang MJJ (1993) A fuzzy multi-criteria decision-making approach for robot selection. Robot Comput Integr Manuf 10(4):267–274. https://doi.org/10.1016/0736-5845(93)90040-Q
Lissenden J, Maley S, Mehta K (2015) An era of appropriate technology: evolutions, oversights and opportunities. J Humanit Eng 3(1):24–35. https://doi.org/10.36479/jhe.v3i1.34
McKay FH, Singh A, Singh S, Good S, Osborne RH (2016) Street vendors in Patna, India: understanding the socio-economic profile, livelihood and hygiene practices. Food Control 70:281–285. https://doi.org/10.1016/j.foodcont.2016.05.061
Naik A (2013) Contextualising urban livelihoods: street vending in India. SSRN Electronic Journal. https://doi.org/10.2139/ssrn.2238589
Prahalad CK (2005) The fortune at the bottom of the pyramid: eradicating poverty through profits. Wharton School Publishing
Reddy AK (1979) Problems in the generation of appropriate technologies. In: Appropriate technologies for third world development. Springer, Berlin, pp 173–189
Rodríguez J, Diehl JC, Christiaans H (2006) Design toolbox for contextualizing products for users in emerging markets. In: IEA2006. Elsevier, Maastricht
Saaty TL (1977) A scaling method for priorities in hierarchical structures. J Math Psychol 15(3):234–281. https://doi.org/10.1016/0022-2496(77)90033-5
Sijali IV, Mwago MG (2011) MoneyMaker pumps: creating wealth in sub-Saharan Africa. In: Chuhan-Pole P, Angwafo M (eds) Yes Africa can: success stories from a dynamic continent. The World Bank, Washington, chap 18, pp 319–336. http://elibrary.worldbank.org/doi/book/10.1596/978-0-8213-8745-0
Teitel S (1978) On the concept of appropriate technology for less industrialized countries. Technol Forecast Soc Change 11(4):349–369. https://doi.org/10.1016/0040-1625(78)90018-5
Triantaphyllou E, Sánchez A (1997) A sensitivity analysis approach for some deterministic multi-criteria decision-making methods. Decis Sci 28(1):151–194. https://doi.org/10.1111/j.1540-5915.1997.tb01306.x
van der Bijl-Brouwer M, Dorst K (2017) Advancing the strategic impact of human-centred design. Design Stud 53:1–23. https://doi.org/10.1016/j.destud.2017.06.003
Wood CD, Lewis PK, Mattson CA (2012) Modular product optimization to alleviate poverty: an irrigation pump case study. In: Volume 3: 38th design automation conference, parts A and B. American Society of Mechanical Engineers, vol 20, pp 455–462. https://doi.org/10.1115/DETC2012-71171
Zhou J, Jiao H, Li J (2017) Providing appropriate technology for emerging markets: case study on China's solar thermal industry. Sustainability (Switzerland) 9(2). https://doi.org/10.3390/su9020178

Chapter 4
Design Analyses of BoP Technologies

In the previous chapter, a strategy was presented to aid in the selection of appropriate technologies for users at the base of the economic pyramid (BoP). The approach is helpful for selecting the best alternative. However, in order to assess the technologies from an engineering design point of view, it is necessary that the design analysis of the technologies is carried out. In this chapter, analyses of two technologies designed for the BoP are presented. The first case focuses on a wooden palanquin (commonly known as a *palki*) used during pilgrimages to Yamunotri *Dham* (Bhat et al. 2019). The structural performance of the existing design is examined to assess its load-bearing capacity and determine whether it is safe and ergonomically viable for use by both the porters and the passengers. The second case involves a treadle pump, a manually operated irrigation device. Given that the pump includes moving parts, both structural and dynamic analyses were conducted. The multibody dynamics modelling incorporated joint clearances to simulate realistic motion. Additionally, since the device is often used for extended periods by a diverse range of users, including women, men, and children, its ergonomic aspects were also considered as part of the study.

4.1 Yamunotri Palanquin (*Palki*)

Yamunotri *dham* is one of the sacred *Char Dham* pilgrimage sites in India. Located at an elevation of approximately 3290 metres, it marks the origin of the Yamuna River and attracts hundreds of thousands of devotees annually. Due to the region's steep and rugged terrain, the final 6.5 kilometres to the shrine must be traversed on foot or with the aid of animals such as mules and horses or manual carriers. One of the most common modes of transport for elderly or mobility-impaired pilgrims is the *palki*, which is a traditional palanquin carried by four porters (Fig. 4.1).

S. Bhat et al., *Technologies for the Base of the Economic Pyramid*, Design Science and Innovation, https://doi.org/10.1007/978-981-95-2721-2_4

Fig. 4.1 A devotee being carried on a *palki* (palanquin) by four porters

These *palkis* are locally fabricated using wood and metal components, based on traditional knowledge passed down through generations (Chakravarthy 2019, 2016). Despite their widespread use, these designs have largely remained unchanged and have not been subjected to systematic engineering analysis. Preliminary observations and interviews with porters suggest that the current designs are suboptimal in terms of structural safety and user comfort. Porters report frequent repairs, discomfort, and strain-related injuries, highlighting the need for ergonomic and structural improvements.

For porters in Yamunotri, the *palki* is a critical livelihood tool, especially during the pilgrimage season. Any failure in design not only affects the safety of the passenger but also imposes financial and physical burdens on the porters. Thus, there is a pressing need to reassess the design through the lens of appropriate technology, prioritising affordability, local manufacturability, material availability, and ergonomic suitability. The goal of this study is to evaluate its current performance under realistic conditions and identify opportunities for redesign.

4.1.1 Current Design of the Palki

Currently, the *palki* is locally manufactured using Pine Wood as the primary material. The design comprises a cushioned wooden chair, a footrest, and two wooden bars fixed to the structure. Vertical wooden bars positioned at the front of the seat provide hand support for the passenger during travel. To enhance structural rigidity, two mild steel (MS) rods are incorporated into the design as reinforcement members. Notably, the frame is slightly inclined to accommodate the natural slope of the terrain, thereby maintaining the passenger in a level and stable orientation during the journey. A 3D model rendered from the inputs of local vendors is shown in Fig. 4.2. This model serves as the basis for the subsequent structural analysis presented in the next section.

Fig. 4.2 CAD model of the *palki*

4.1.2 Structural Assessment Methodology

This section describes the general method used to undertake the structural assessment of the *palki*. The approach integrates field observation, dimensional and ergonomic suitability, modelling, and analysis.

4.1.2.1 Field Observation and Data Acquisition

Measurements of the *palki* were obtained from the local manufacturers and validated through field visits. Interviews with porters and passengers revealed practical concerns such as joint instability, shoulder pain, and imbalance during the travel. Dimensional information was also obtained from the field. The *palki* has a typical length of 2.4 metres, and wooden logs of approximately 1 metre length are mounted on both ends for lifting. The total structure weighs between 18 and 20 kg.

4.1.2.2 Ergonomics

Some of the key ergonomic parameters were noted during the study. This is important mainly to assess user comfort. The minimum seat width recommended for the Indian population is 394 mm (Chakrabarti 1997). The seat width in the current design is 530 mm, suggesting sufficient lateral space. However, the seat height was only 195 mm, which is significantly lower than the recommended 435 mm.

4.1.3 Modelling and Analysis

The *palki* is made of longleaf pine, a variety found commonly in the Uttarakhand region, where the shrine is situated. While wood has complex orthotropic behaviour due to grain direction Green et al. (1999), a simplified material model was adopted to approximate its load-bearing characteristics (Olorunnisola 2018).

Aligning with the field observation, a total passenger load of 75 kg distributed across the seat and footrest regions was considered. Four-point supports were applied at the ends of the lifting logs to represent manual lifting by porters. Inclined ground conditions were replicated to reflect the typical operating slope encountered in the Yamunotri terrain. These inputs were derived from direct observation of porter postures and terrain mapping.

A static structural analysis was conducted to identify critical stress concentrations, joint vulnerability zones, and load path inefficiencies. It is to be noted that the goal of the analysis was not to determine precise stress values, but to identify vulnerable regions in the structure that were at risk of material and structural failure.

4.1.4 Results and Discussions

The structural simulation of the *palki* revealed several critical insights related to safety, usability, and long-term reliability of the structure.

A high-stress region was observed in the MS rods, which acted as a primary load-bearing element. While the rods improved the load-carrying capacity, they did not effectively distribute stress across the region. Higher stress was also observed in the junctions between wooden bars and the vertical support. These areas are vulnerable due to the load from passenger weight and dynamic motion during the travel. The analysis results were also consistent with field observations, where porters had used reinforcements in the form of ropes and wires in the regions of higher stress concentration. The inclination of the frame maintains passenger posture parallel to the ground. However, this design shifts the load distribution towards one set of porters depending on the direction of travel.

The analysis shows that the current design is only marginally safe. Moreover, it is noteworthy that the properties of wood vary a lot based on factors such as moisture, quality of wood, and cutting direction. This mandates a higher factor of safety for the *palki* design. While more aspects of the usage condition, such as dynamic loading and structural stability, need to be studied, preliminary analysis reveals several opportunities for design improvement. Areas such as joints, reinforcements, and ergonomics can be readily improved based on the results of this study.

This case highlights how even a simple, low-cost structure can benefit from basic engineering analysis. Effective and systematic use of field insights and collaboration with relevant stakeholders can help greatly in problem identification and validation.

4.2 Treadle Pump

The treadle pump by RuTAG IIT Delhi is a demand-driven intervention that originated in a village called Deoria in Uttar Pradesh, India. In the context of BoP, where maintenance facilities are limited and users may include women, children, and the elderly, it becomes important to assess the structural integrity, efficiency, and user comfort. To assess the appropriateness of the design, three interrelated analyses were conducted:

1. Static Structural Analysis: to evaluate the ability of the frame and components to withstand various usage loads
2. Joint Clearance Analysis: to study the effects of mechanical play in linkages, which are commonly found in devices fabricated in low-resource settings
3. Ergonomic Analysis: to determine whether the design accommodates a diverse user base in terms of comfort, fatigue, and biomechanical efficiency during prolonged operation

In addition to revealing opportunities for design refinements, the analyses also serve as a model for evaluating similar other BoP technologies.

4.2.1 Structural Analysis

The important structural elements of the treadle pump include the lever, the treadle, the piston, and the cylinder. In the RuTAG IITD treadle pump, the assembly of levers and the rocker link (Fig. 4.3) can be considered as a six-link mechanism.

4.2.1.1 Working Principle

The RuTAG IITD treadle pump is a suction pump. A suction-type pump works in two strokes: suction stroke and return stroke (Fig. 4.4). During the suction stroke, the user presses the lever, which causes the piston to rise upwards in the cylinder. This generates a suction pressure in the cylinder, causing the water to flow in from the inlet pipe. During the return stroke, the piston moves down, causing the piston valve to open, and water gets collected over the piston washer. This water is delivered through the channel during the next suction stroke. In the case of the RuTAG IITD

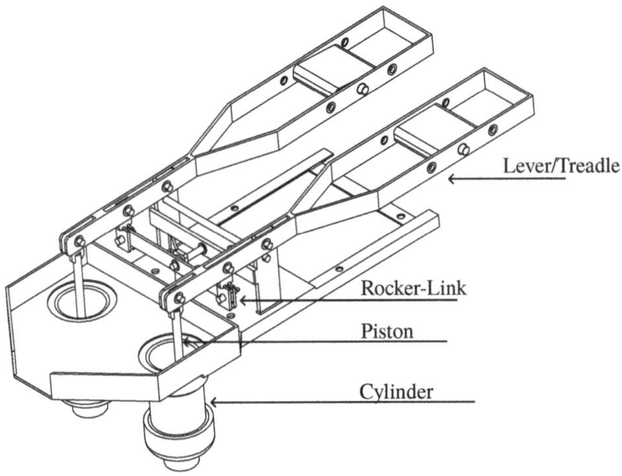

Fig. 4.3 Mechanism of the RuTAG IITD treadle pump

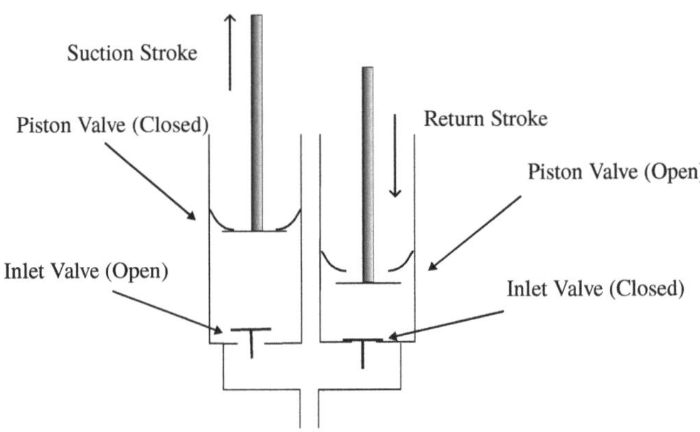

Fig. 4.4 Working principle of the treadle pump

treadle pump, the reciprocating motion is achieved through the rocker linkages (Fig. 4.3). When a lever is pressed, the rocker link provides a reaction force to the opposite lever, causing it to move up. The forces acting on the lever during both strokes are shown in Fig. 4.5.

Here, F_f is the foot force applied by the user, F_r is the reaction due to forces on the opposite treadle, F_s is the suction force generated during the suction stroke, and F_d is the drag force during the return stroke. If ρ is the density of water, a is the cross-sectional area of the cylinder, and h is the depth of the water source, the suction force can be calculated as

$$F_s = \rho g a h \tag{4.1}$$

Fig. 4.5 Forces acting on the treadle. (**a**) Suction stroke. (**b**) Return stroke

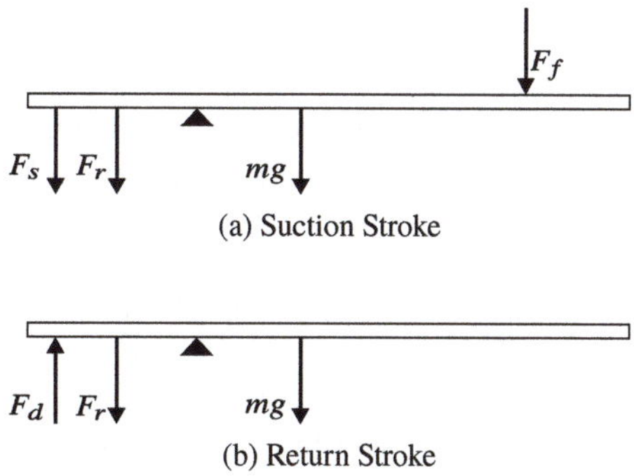

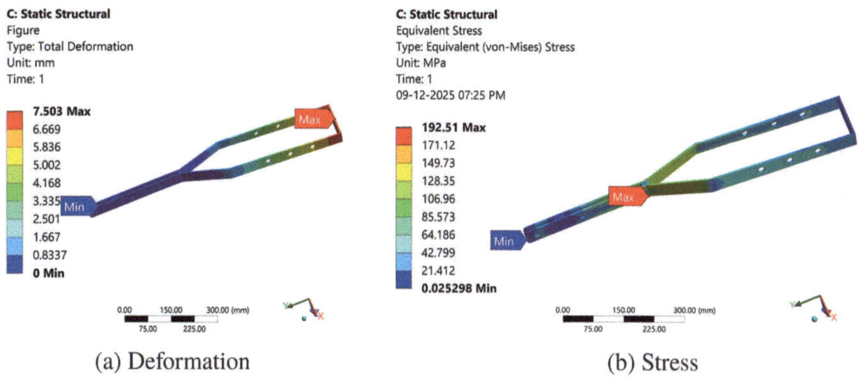

Fig. 4.6 FE analysis of the treadle. (**a**) Deformation. (**b**) Stress

Similarly, for the return stroke, drag force can be calculated using Eq. 4.2, where C_d is the coefficient of drag, and v is the linear velocity of the piston.

$$F_d = \frac{C_d \rho v^2 a}{2} \tag{4.2}$$

The foot force F_f and reaction forces F_r can be calculated using moment balance. It can be observed that the foot force is dependent on the depth of the water source and the treadling speed. The results of the FE analysis done on the treadle are shown in Fig. 4.6. For this analysis, the piston end of the lever is fixed, and a static load of 600 N (approximate average weight of the user) was considered. It can be seen that a maximum deflection of 7.5 mm was observed along with a stress of 192.5 MPa, which is within the yield stress limit of mild steel. Because of the small and limited motion of the treadle (±15°), only static FE analysis was considered. More insights could be found through the multi-body dynamic modelling discussed in the next section.

4.2.2 Dynamic Analysis

It was seen in the previous section that the design of the treadle pump is found to be structurally safe if static loads are considered. Since the treadles are moving parts, it is essential to understand their dynamic behaviour. The dynamics of the treadle pump is modelled here using the DeNOC formulation (Saha 1999). In the case of RuTAG IITD treadle pump, two three-link mechanisms are coupled by a rocker linkage. If we apply Grubler's criterion to this system, the degree of freedom comes out to be zero. The limited reciprocating motion observed can thus be attributed to joint clearances and the elasticity of the leather pistons used. The system can be represented as an equivalent four-link mechanism as shown in Fig. 4.7. For each stroke, the treadle makes an angular displacement of 15° from the zero position.

In order to write the equation of motion of the system, the closed-loop four-link mechanism can be represented as two subsystems shown in Fig. 4.8.
The equations of motion can then be derived (Chaudhary and Saha 2008) in the form of

$$\begin{bmatrix} \mathbf{I}_{\mathrm{I}} & \mathbf{O} & -\mathbf{J}_{\mathrm{I}}^T \\ \mathbf{O} & \mathbf{I}_{\mathrm{II}} & -\mathbf{J}_{\mathrm{II}}^T \\ -\mathbf{J}_{\mathrm{I}} & -\mathbf{J}_{\mathrm{II}} & \mathbf{O} \end{bmatrix} \begin{bmatrix} \ddot{\boldsymbol{\theta}}_{\mathrm{I}} \\ \ddot{\boldsymbol{\theta}}_{\mathrm{II}} \\ \boldsymbol{\lambda} \end{bmatrix} = \begin{bmatrix} \boldsymbol{\tau}_{\mathrm{I}} - \mathbf{h}_{\mathrm{I}} \\ \boldsymbol{\tau}_{\mathrm{II}} - \mathbf{h}_{\mathrm{II}} \\ \dot{\mathbf{J}}_{\mathrm{I}}\dot{\boldsymbol{\theta}}_{\mathrm{I}} + \dot{\mathbf{J}}_{\mathrm{II}}\dot{\boldsymbol{\theta}}_{\mathrm{II}} \end{bmatrix} \tag{4.3}$$

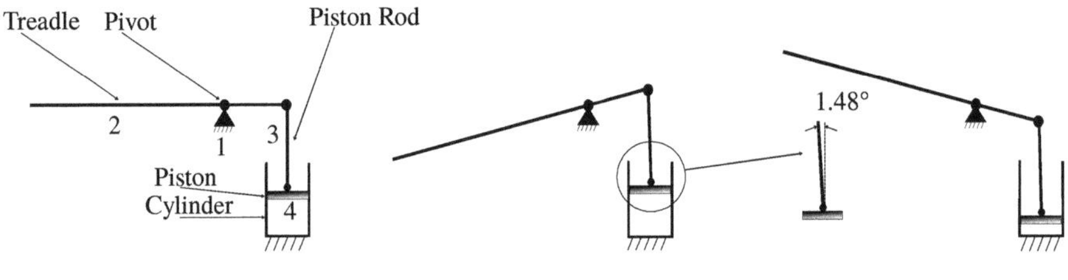

Fig. 4.7 Treadle pump as a four-link mechanism

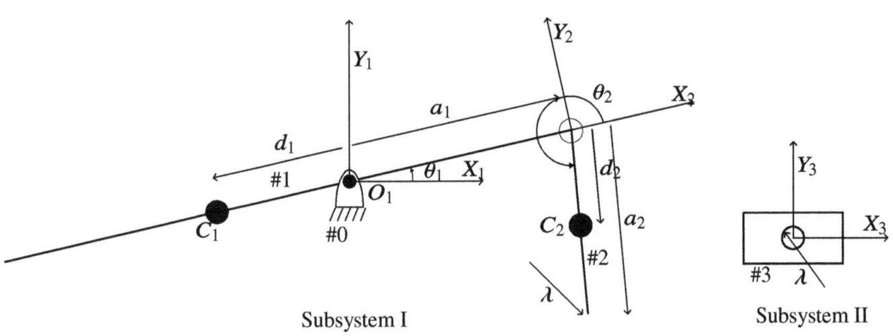

Fig. 4.8 Representation of four-link mechanism as two subsystems

where $\mathbf{I}_{\rm I}$ and $\mathbf{I}_{\rm II}$ are the generalised inertia matrices of subsystems I and II, respectively, $\mathbf{J}_{\rm I}$ and $\mathbf{J}_{\rm II}$ are the constraint Jacobians, $\boldsymbol{\theta}_{\rm I}$ and $\boldsymbol{\theta}_{\rm II}$ are the vectors containing generalised coordinates, $\boldsymbol{\lambda}$ is the vector of Lagrange multipliers, $\boldsymbol{\tau}_{\rm I}$ and $\boldsymbol{\tau}_{\rm II}$ are the generalised force vectors, and $\mathbf{h}_{\rm I}$ and $\mathbf{h}_{\rm II}$ are the vectors of convective inertia terms. The derivation of the matrices (Chaudhary and Saha 2008; Saha 2014) can be found in Appendix B.

4.2.2.1 Clearances

The equations of motion shown in Eq. 4.3 assume perfect revolute joints in the mechanism. However, it is known that, in reality, joint clearances are required in order to facilitate the movement of the links. Moreover, in BoP settings, it is observed that the required tolerances are not achieved due to the limitations of available manufacturing facilities. It is thus essential that the effects of joint clearances on the performance of the devices are studied before a decision is made while implementing it in the field.

For this study, the joint clearance between two links is modelled as two colliding bodies, namely a pair of journal and bearing, and a multibody dynamic approach is considered. Elastic contact is assumed between the journal and the bearing, where deformations and penetrations are considered. In a revolute joint, the interactions between the journal and the bearing are represented by force constraints. The forces generated during the impact are dependent on the penetration depth, velocities of the bodies, and the material properties of the journal and the bearing.

Figure 4.9 shows the impact between bearing i and journal j. With respect to the global frame XY, the vectors $\boldsymbol{r}_i^P$ and $\boldsymbol{r}_j^P$ represent the position of bearing and journal centres P_i and P_j, respectively. Similarly, $\boldsymbol{r}_i^Q$ and $\boldsymbol{r}_j^Q$ represent the position of contact points of bearing and journal, and δ represents the penetration depth. The normal contact force F_N can be obtained using Hertz's contact law as

$$F_N = K\delta^n \tag{4.4}$$

where δ is the penetration depth, the exponent n is taken as 1.5 for spherical contact between two metallic surfaces, and the generalised stiffness constant K is given as (Hertz 1896)

$$K = \frac{2}{3h}\left(\frac{R_i R_j}{R_i + R_j}\right)^{1/2} \tag{4.5}$$

where R_i and R_j are the radii of the two links, as shown in Fig. 4.9. Moreover, if ν and E are Poisson's coefficient and Young's modulus of the material of the links, h can be given by Hertz (1896)

$$h = \frac{1 - \nu^2}{E} \tag{4.6}$$

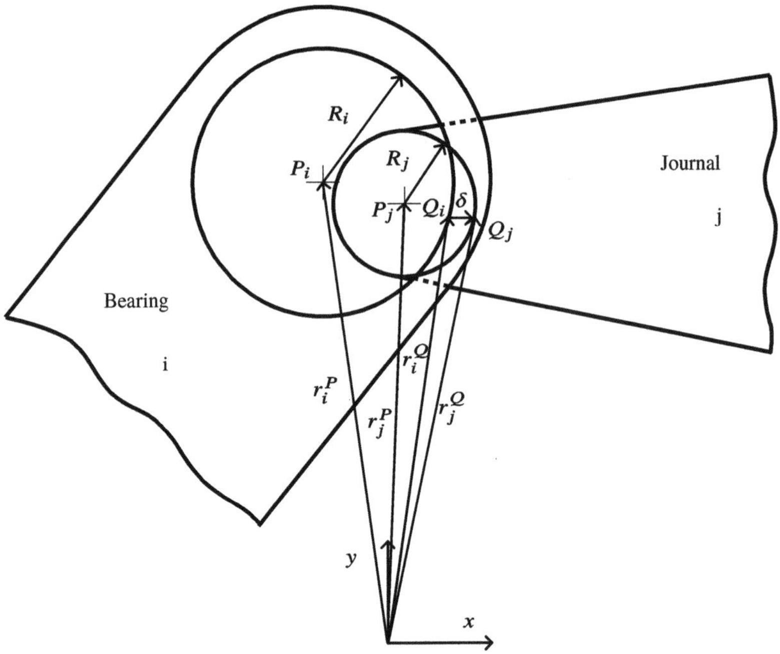

Fig. 4.9 Impact of journal and bearing

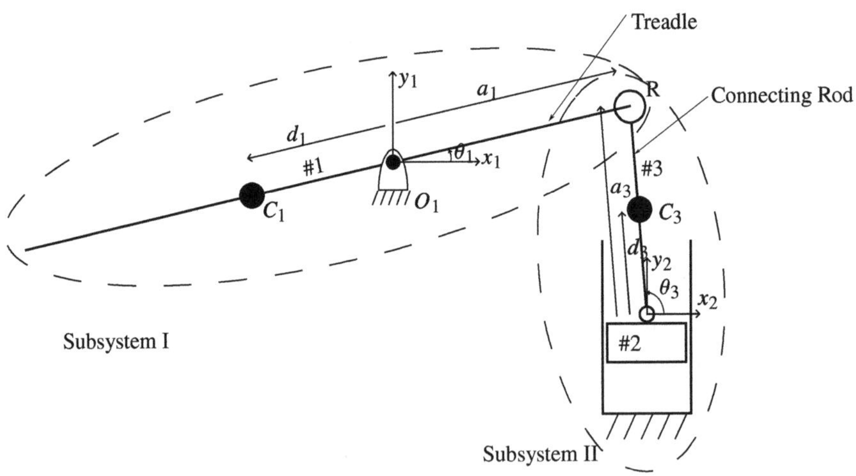

Fig. 4.10 Clearance between treadle and connecting rod

If clearance is considered in the revolute joint R of the treadle pump system as shown in Fig. 4.10, it introduces two more degrees of freedom in the system since θ_3, and slider motion becomes independent of θ_1. The system is then constrained by the contact forces generated during the impact between the journal of link #1 and

Table 4.1 Properties of the links

Link	Mass (kg)	Length (m)	Centre of mass location* (m)	Moment of Inertia (kgm^2)
Treadle	2.88	0.82	− 0.2	0.192
Piston	0.284	–	0	–
Connecting rod	0.314	0.112	0.23	1.29×10^{-3}

*Centre of mass is measured from the origin of the link

the bearing of link #3. To apply the concept of subsystem approach (Chaudhary and Saha 2008), the system is divided into two subsystems I and II, where subsystem I is the single link treadle constrained at the pivot O_1, and subsystem II consists of the piston and connecting rod, which forms a PR mechanism with one prismatic and one revolute joint. Following the usual notations (Chaudhary and Saha 2008), the equations of motion of the two subsystems are written as

$$\begin{bmatrix} \mathbf{I}_{\mathrm{I}} & \mathbf{O} \\ \mathbf{O} & \mathbf{I}_{\mathrm{II}} \end{bmatrix} \begin{bmatrix} \ddot{\boldsymbol{\theta}}_{\mathrm{I}} \\ \ddot{\boldsymbol{\theta}}_{\mathrm{II}} \end{bmatrix} = \begin{bmatrix} \boldsymbol{\tau}_{\mathrm{I}} + \boldsymbol{\gamma}_{\mathrm{I}} \\ \boldsymbol{\tau}_{\mathrm{II}} - \mathbf{h}_{\mathrm{II}} + \boldsymbol{\gamma}_{\mathrm{II}} \end{bmatrix} \quad (4.7)$$

where $\boldsymbol{\gamma}$ represents the vector of gravity terms. The expanded terms of the matrices (Chaudhary and Saha 2008) and the derivations of the same can be found in Appendix C. The link properties are shown in Table 4.1. The material properties of mild steel ($E = 206$ GPa and $\nu = 0.3$) were considered for the analysis. Further, a radial clearance of 1 mm was considered for the analysis. The motion of the treadle was simulated for a single power stroke, that is, for a complete motion of the piston from the lower end of the cylinder to the top end. The corresponding motion of the treadle is $\theta_1 = -15°$ to $+15°$. Simulations were performed using Eq. 4.7, whose results are shown in Fig. 4.11. Note the motion of the journal (lever) inside the bearing (connecting rod) in Fig. 4.11a. The multiple collisions have a direct effect on the usability of the treadle pump. It can be observed in the variations of the piston's speed shown in Fig. 4.11b. The jerkiness in the motion not only impacts the user's experience but also causes damage to the joints, thereby affecting the output of the pump and its efficiency.

4.2.2.2 Optimisation

Due to the non-availability of precision manufacturing facilities, maintaining a high degree of tolerance is challenging in a low-resource setting. In such cases, modifying mass and inertia properties can be an effective way to improve the performance of the devices. By redistributing the mass across the links, i.e., redesigning the link's shape, material, etc., one can alter the dynamic responses of the system. In the present case, one can improve the performance of the treadle pump by minimising the impact forces acting between the journal and the bearing. This can be achieved

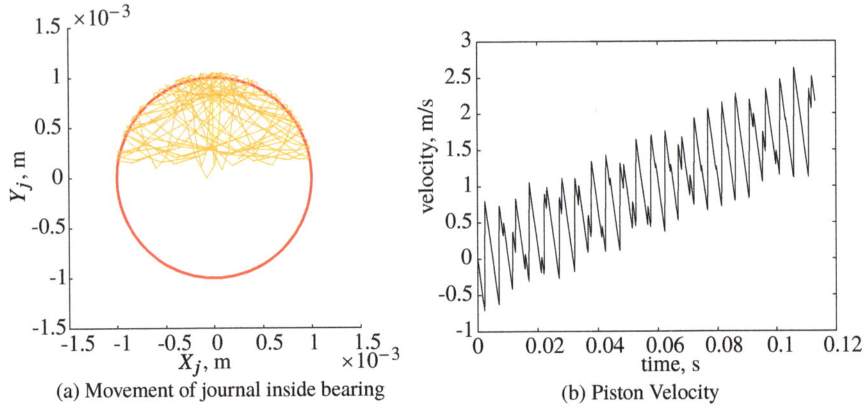

Fig. 4.11 Effect of joint clearance on the treadle pump. (**a**) Movement of journal inside bearing/ (**b**) Piston velocity

by maintaining the contact between the journal and the bearing throughout the duration of motion (Varedi et al. 2015). If δ_x and δ_y represent the relative distances between the x and y coordinates of the journal and bearing centres, respectively, the eccentricity e can be obtained as

$$e = \sqrt{\delta_x^2 + \delta_y^2} \tag{4.8}$$

The relative penetration δ of Fig. 4.9 can then be obtained as

$$\delta = e - c \tag{4.9}$$

where c is the radial clearance defined as the difference between the radii of the journal and the bearing.

If the solution to the set of equations in Eq. 4.7 is obtained in n time steps, the objective function F to be minimised can be written as

$$F = \frac{1}{n}\sqrt{\sum_{i=1}^{n} \delta_i^2} \tag{4.10}$$

where δ_i is the relative penetration for the ith time step. The design variables considered for this optimisation problem are the mass, mass centre location, and the mass moment of inertia of the treadle and the connecting rod. The design variables and the upper and lower limits considered for the optimisation are shown in Table 4.2. The values were based on practical limitations such as space availability. Particle Swarm Optimisation (Kennedy and Eberhart 1995) was used to solve the problem using MATLAB. The optimal values obtained are shown in Table 4.3.

Table 4.2 Design variables for the optimisation

	$m_1(kg)$	$d_1(m)$	$I_{1z}(kg-m^2)$	$m_3(kg)$	$d_3(m)$	$I_{3z}(kg-m^2)$
Lower bound	1	− 0.5	0.05	0.1	0.05	1×10^{-4}
Upper bound	5	− 0.1	4	3	0.2	5×10^{-3}

Table 4.3 Optimal values of the design variables

Design variable	Original value	Optimal value
$m_1(kg)$	2.88	1.345
$d_1(m)$	− 0.2	− 0.3265
$I_{1z}(kg-m^2)$	0.192	0.1301
$m_3(kg)$	0.314	1.6426
$d_3(m)$	0.112	0.05
$I_{3z}(kg-m^2)$	1.29×10^{-3}	0.0013

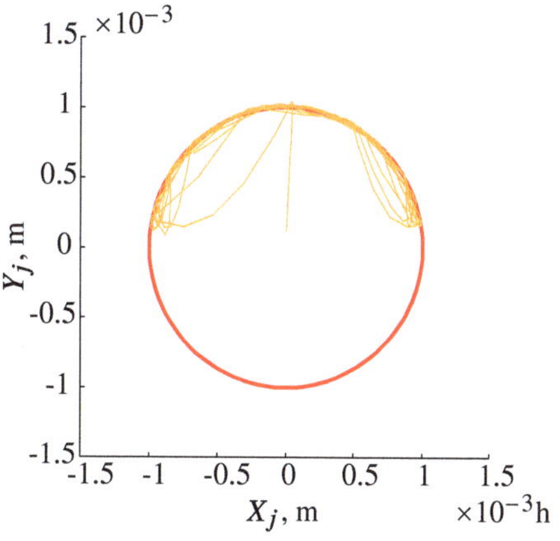

Fig. 4.12 Journal movement for optimised link properties

The resultant motion is shown in Fig. 4.12. It can be observed how the journal slides along the edge of the bearing, thus minimising the effects caused due to multiple collisions. The comparison of velocities can be seen in Fig. 4.13. It can be observed in the optimised configuration that the jerkiness caused due to sudden changes in angular and linear velocities of the links has been reduced, and the motion is smoother.

It is to be noted that, for optimisation, only mass and inertia properties were considered as design variables. The optimal values thus obtained are limited to this case study. Moreover, techniques such as shape optimisation and synthesis can be used to realise the optimal shape and mass-inertia properties. Additionally, constraints such as stress, factor of safety, and treadling forces can also be used for optimisation.

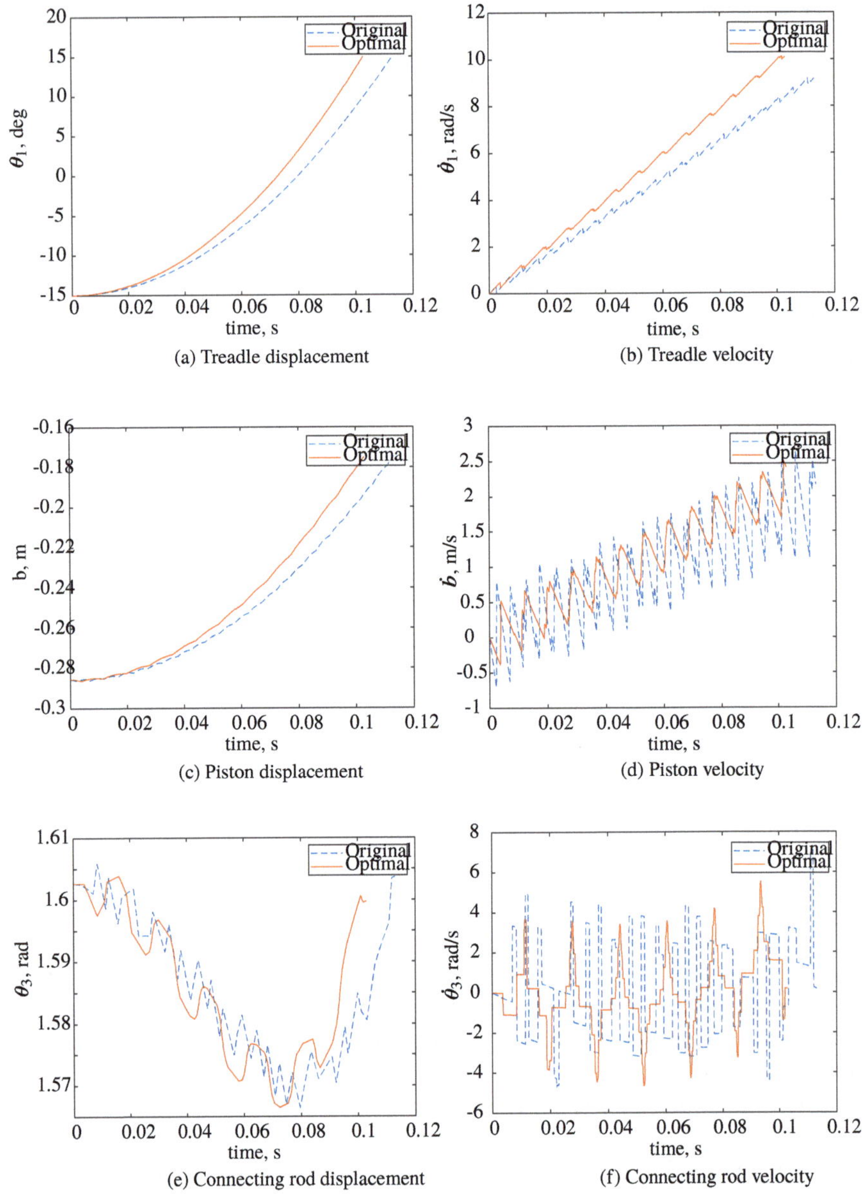

Fig. 4.13 Displacement and velocities of the links of treadle pump. (**a**) Treadle displacement. (**b**) Treadle velocity. (**c**) Piston displacement. (**d**) Piston velocity. (**e**) Connecting rod displacement. (**f**) Connecting rod velocity

4.2.3 Ergonomic Analysis

In general, *ergonomics* can be defined as the science of designing products such that users interact with them efficiently. Sometimes, it is also known as *human factors*. Sanders and McCormick (1987) defined two objectives of ergonomics, namely to enhance the effectiveness and efficiency of the work and to enhance certain desirable human values such as safety, comfort, and acceptance. Some of the factors that play a role in ergonomics include body posture and movement, environmental factors, information and operation, and work organisation (Dul and Weerdmeester 2003). See Table 4.4.

Many researchers have highlighted the importance of considering ergonomic factors in the design for the BoP. For instance, Sethi (1989) wrote about the ergonomic considerations while designing a prosthetic limb in the Indian context. He wrote about how the cultural and habitual practices concerning sitting on the chair or the floor are different in the Indian context compared to the West. Poor ergonomics have also been linked to the unacceptability of the design by the users (Lecomte et al. 2013; Jagtap et al. 2014; Bergmann et al. 2015). The authors also stated that despite the failures caused, designers often tend to overlook the ergonomic factors while designing for the BoP. It can also be seen how ergonomics plays a crucial role in human-centred design (van der Bijl-Brouwer and Dorst 2017) and design for sustainability (Nadadur and Parkinson 2013). Fisher (2006) lists *ergonomic design and safety* as one of the design criteria for the kickstart pumps. In the design analysis of BoP technologies, ergonomic factors were considered in the case of medical devices (Agarwal et al. 2021), agricultural equipment (Bautista et al. 2017), and irrigation pumps (Pereira et al. 2006; McComb et al. 2014). McComb et al. (2016) have considered ergonomic factors such as step height, operating force, and treadle width in the optimisation study done for IDE-style treadle pumps. Pereira et al. (2006) looked into the dimensional and structural aspects of a treadle pump design. The author highlighted that since the pump is operated by men, women, and children for long periods, comfort is essential. The author recommended the usage of biomechanical models and the concept of muscle fatigue to analyse the human-machine interaction in order to increase productivity and comfort.

Table 4.4 Some of the ergonomics factors to be considered in design

Posture and movement	Information and operation	Environment	Work organisation
Sitting	Visual	Noise	Tasks
Standing	Hearing	Vibration	Jobs
Lifting	Touch and haptics	Illumination	Work time
Pulling	Controls	Climate	
Pushing		Chemical substances	

(a) Sitting position (b) Standing position

Fig. 4.14 Treadle pump operating positions. (**a**) Sitting position. (**b**) Standing position

Treadle pumps are used as a source of primary irrigation in many places. As a result, the user operates the pump for a long duration throughout the day. Hence, the analysis of human-machine interaction becomes crucial. It can be observed that the treadle pump operation involves repetitive motions of the legs (Fig. 4.14). The multi-body dynamics approach can be adapted to analyse the ergonomics of the treadle pump. Joint torques were obtained corresponding to the treadle pump motion using inverse dynamics. The concept of muscle fatigue was then considered to estimate the duration for which a user can operate the pump continuously. Based on the obtained results, an optimised configuration of the treadle pump is also suggested.

4.2.3.1 Biomechanical Modelling

To simulate human-machine interaction, the human body may be represented using biomechanical models. These models are used in applications such as vehicle crash analysis (Silva et al. 1997; Portal et al. 2006), motion simulation and gait analysis (Blajer et al. 2007; De Groote et al. 2009; Ambrosio 2013), orthopaedics and prosthesis development (Reinbolt et al. 2005; Copilusi et al. 2019), and ergonomic studies (Pereira et al. 2006; Castellone et al. 2017). Biomechanical models with varying degrees of freedom (DoF) were developed based on the application. For instance, Silva et al. (1997) developed a 29 DoF model with 12 rigid bodies and 11 kinematic joints as depicted in Fig. 4.15. The mass and inertial properties of

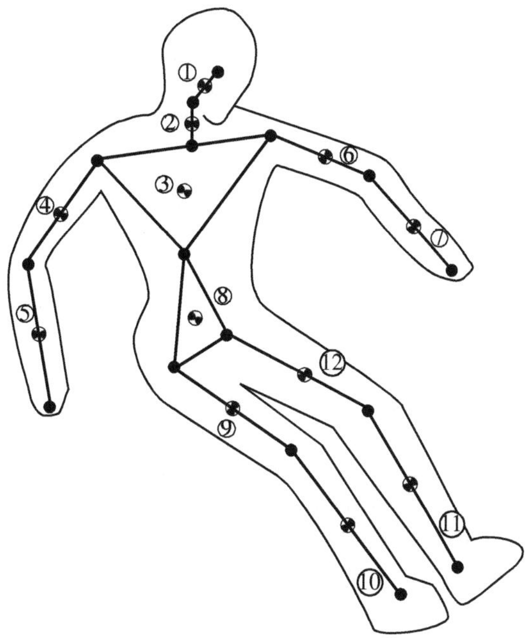

Fig. 4.15 Twelve-link model of a human body (Silva et al. 1997)

the rigid bodies were assigned based on the anthropometric data available in the literature. Ayoub (1998) used a 2D five-link rigid body model to simulate lifting tasks. Ning et al. (2004) used a 14-link model to track human motion.

The joint forces and torques for a given motion can be obtained through the biomechanical models using inverse dynamics. However, in a musculoskeletal system, the muscles associated with the concerned motion, sometimes modelled as linear actuators, are often more than the number of degrees of freedom. The calculation of muscle forces thus becomes a problem with redundancy. Optimisation techniques are used to solve such problems (Silva and Ambrósio 2003). Here, it is important that the constraints considered for the optimisation problem reflect the actual physiological response of the human body. Some authors have used muscle activation models to replicate the response of the human body (Zajac 1989; Anderson and Pandy 2001). One of the most popular models is the muscle activation-contraction model developed by Hill (1938, 1953).

It is to be noted that, unlike mechanical actuators, the capacity of human muscles to exert force reduces over time. This phenomenon is known as muscle fatigue. The actual response of the exertable muscle force is dependent on various physiological and neurological factors. There are some empirical models, such as the Rohmert curves, which predict the muscle fatigue behaviour over time (Rohmert 1960). Theoretical models were also developed, which incorporate the biological and physical behaviour of the muscles (Ding et al. 2000; Xia and Law 2008). Liu et al. (2002) used compartment theory, which modelled muscle fibres in three states, namely resting, activated, and fatigued. This model uses the fatigue factor, resting factor, and the concept of motor units to model the fatigue and recovery behaviour of the muscles. Dynamic muscle fatigue model was introduced by Ma et al. (2009). The authors defined *fatigue index* as a function of maximum voluntary contraction (maximum capacity of the muscle), current exertable force, and the current load on the muscle. Ma et al. (2012) later also showed that the same relationship also holds true with joint torques. It was seen that muscle fatigue is proportional to the joint torque and inversely proportional to the muscle torque capacity.

Note that the treadle pump by RuTAG IITD can be operated in two positions, namely sitting and standing, as shown in Fig. 4.14. Researchers have used several multibody models of the human body for various applications such as crash analysis, motion simulation, and ergonomic analysis. For instance, Silva et al. (1997) used a 12-link rigid body model for crash analysis. Here, the lower limbs of the human body are modelled to analyse the ergonomic performance. The mass and inertia properties for each link were used from the available data (Silva et al. 1997), and inverse dynamics was performed to obtain the joint torques for various preset motions. Since the treadle pump used in this analysis is primarily used in India, the link lengths were as per the anthropometric data for Indian adult males Chakrabarti (1997).

4.2.3.2 Sitting Position

In the sitting position, since the motion of the upper body is limited, only the lower limbs are considered for the analysis. If the treadle pump and the lower limbs of the human body are considered to be a single mechanism, a simplified 2D model can be represented as shown in Fig. 4.16a. This is an equivalent four-bar mechanism. In this mechanism, link 1 represents the lever, and links 2 and 3 are the lower and

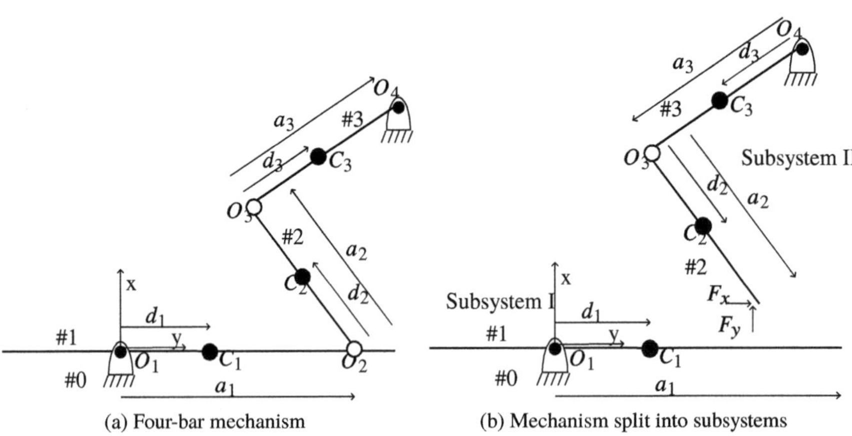

Fig. 4.16 Mechanism for sitting position. (**a**) Four-bar mechanism. (**b**) Mechanism split into subsystems

Table 4.5 Properties of links

Link no.	Length (m)	d (m)	Mass (kg)	I_{xx} $(kg-m^2)$	I_{yy} $(kg-m^2)$	I_{zz} $(kg-m^2)$
1	0.85	0.2	2.88	5.64×10^{-3}	0.197	0.192
2	0.467	0.23	4.808	3.33×10^{-4}	1.48×10^{-4}	3.42×10^{-4}
3	0.434	0.215	9.843	1.25×10^{-4}	3.43×10^{-4}	4.13×10^{-4}

upper legs, respectively. The combined mechanism is a 1-DoF system. However, in reality, there are two actuators in terms of the hip and knee joints. This makes it a system with redundant driving torques. In order to address this, the lever and the lower limbs are considered two separate systems, as shown in Fig. 4.16b. Inverse kinematics is performed on the combined mechanism to obtain the joint angles. Foot force F_f was calculated using the joint angle information for O_1 of subsystem I obtained from the inverse kinematics. The torques at the hip and knee joints were calculated recursively from the two-link mechanism of subsystem II. The foot force obtained from subsystem I acts as an external force for subsystem II. The mass-inertia properties and the centre of mass locations of the leg were obtained from Silva et al. (1997). The properties of the links are shown in Table 4.5.

It is to be observed that the driving torque and the force required to operate the pump depend on the depth of water and the rate of treadling. For the inverse dynamics performed here, the water is assumed to be at a depth of 3 m and the treadling rate to be 30 strokes per minute. Using DeNOC formulation, the equations of motion, similar to Eq. 4.7, were derived for the closed-loop system, which are given below:

$$\begin{bmatrix} \mathbf{I} & -\mathbf{J}^T \\ -\mathbf{J} & \mathbf{O} \end{bmatrix} \begin{bmatrix} \ddot{\boldsymbol{\theta}} \\ \boldsymbol{\lambda} \end{bmatrix} = \begin{bmatrix} \boldsymbol{\tau} - \mathbf{h} + \boldsymbol{\gamma} \\ \dot{\mathbf{J}}\dot{\boldsymbol{\theta}} \end{bmatrix} \tag{4.11}$$

The joint torques are obtained by performing inverse dynamics (Chaudhary and Saha 2008). The methodology is shown in Appendix D. Joint torques during the power and return strokes are shown in Fig. 4.17.

4.2.3.3 Standing Position

In the standing position, the position of the hip cannot be determined from the treadle pump configuration since the seat is not used by the operator. In order to determine the extreme angles of the joints, an AI-based Python tool *MediaPipe Pose* was used (Bazarevsky et al. 2020). This tool identifies predetermined landmarks in a human body from a video or an image. The landmarks include major joints such as the ankle, hip, knee, shoulders, elbow, and wrist. For this case, the code was modified to calculate the joint angles from the identified landmarks. Two images showing the user in the extreme positions of power and return stroke were analysed, and the angles were calculated as shown in Fig. 4.18. In the figure, the red dots

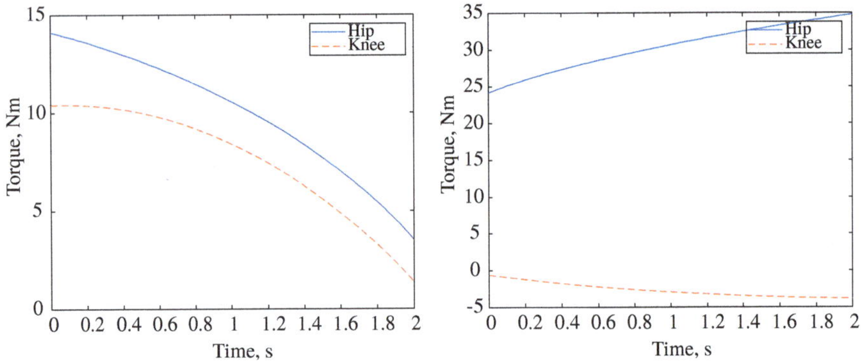

(a) Power Stroke (b) Return Stroke

Fig. 4.17 Joint torques for sitting position. (**a**) Power stroke (**b**) Return stroke

(a) Power Stroke (b) Return Stroke

Fig. 4.18 Joint angles for standing position. (**a**) Power stroke. (**b**) Return stroke

indicate the landmarks identified by the tool, and the green numbers are the angles between the two links. The joint torques were then calculated using the same procedure as explained in Sect. 4.2.3.2. The joint torques for the power and return stroke in the standing position are shown in Fig. 4.19.

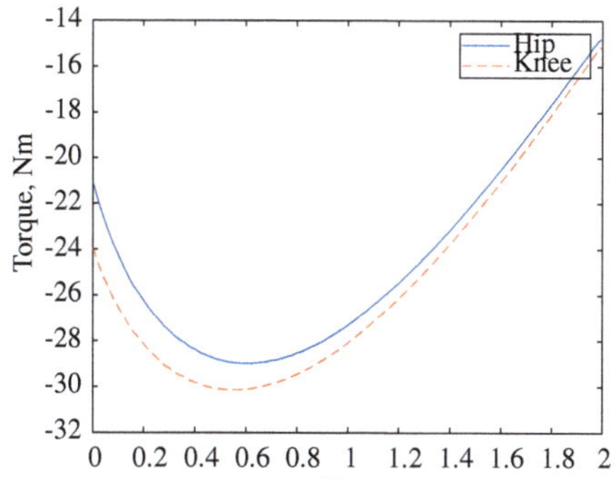

(a) Power Stroke

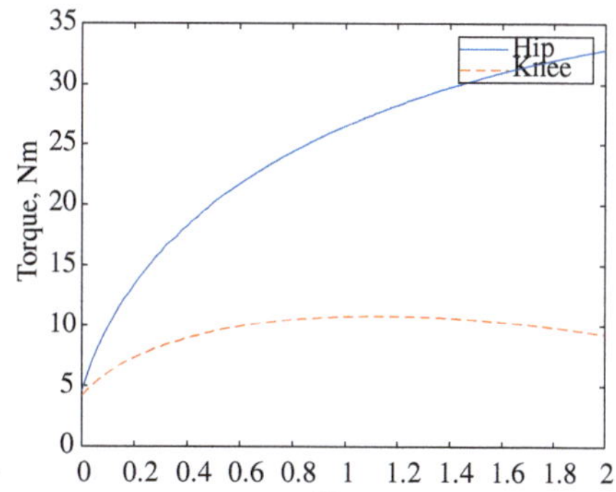

(b) Return Stroke

Fig. 4.19 Joint torques for standing position. (**a**) Power stroke (**b**) Return stroke

4.2.3.4 Muscle Fatigue

To study the effects of repetitive motion on the human body, the concept of muscle fatigue was used. This gives an idea of the degradation of muscle capacity while operating the treadle pump. It can provide an estimate of the maximum duration of the operation. Muscles are the actuators of the human body joints, and modelling their exact behaviour involves complex physiological and neurological factors. Moreover, since the number of actuators is more than the degrees of freedom, the calculation of muscle forces becomes a problem with redundancy. However, simpler models are available that can estimate the muscle activation patterns under various loading conditions (Silva et al. 1997; Ambrosio 2013).

In the case of the treadle pumps, one is more interested in the behaviour of the muscle under dynamic loading conditions. Theoretical and empirical models are available in the literature to evaluate muscle fatigue. Most of the models consider three states of the muscles, namely activation, fatigue, and recovery. In this work, the recovery state is not considered since the objective is to optimise the treadle pump's configuration for the maximum duration of continuous operation. The model proposed by Ma et al. (2009) has been used in this study. The model takes into account the muscle force history and the Maximum Voluntary Contraction (MVC), which is the maximum exertable force by a particular muscle. The fatigue index is then defined as

$$\frac{dU(t)}{dt} = \frac{\text{MVC}}{F_{\text{cem}}(t)} \frac{F_{\text{load}}(t)}{F_{\text{cem}}(t)} \tag{4.12}$$

Here, U is the fatigue index, $F_{\text{cem}}(t)$ is the current exertable maximum force, i.e., the current capacity of the muscle, and $F_{\text{load}}(t)$ is the current external load on the muscle. Further, the rate of change of the current capacity of the muscle is given by

$$\frac{\mathrm{d}F_{\mathrm{cem}}(t)}{\mathrm{d}t} = -k\frac{F_{\mathrm{cem}}(t)}{\mathrm{MVC}}F_{\mathrm{load}}(t) \tag{4.13}$$

where k is a constant and represents fatigue rate with the unit of min^{-1}. The value depends on various factors, such as the energy level of the user and environmental conditions. In ideal conditions, $k = 1/\mathrm{min}$. Ma et al. (2012) later showed that the same relation holds true even for joint torques. The joint-level relationship between the current capacity of the muscle and the current joint torque can be written similar to Eq. 4.13 as

$$\frac{\mathrm{d}\tau_{\mathrm{cem}}(t)}{\mathrm{d}t} = -k\frac{\tau_{\mathrm{cem}}(t)}{\mathrm{MVC}}\tau_{\mathrm{joint}}(t) \tag{4.14}$$

It can be observed that the current capacity of the muscle decreases over time due to the external torque on the joint. The rate of decrease also increases with increased external torque. If the joint is undergoing repeated motions, the muscle is said to be fatigued when the current capacity of the muscle becomes lower than the torque required to perform the motion.

For treadle pump, when the capacity of leg muscles becomes lower than the torque required to draw water from the ground, the user is said to be fatigued. The time duration to reach the minimum torque from the maximum capacity is known as endurance. Here, the instantaneous joint torques can be calculated by using the equations of motion as explained in Sects. 4.2.3.2 and 4.2.3.3. Further, for the initial conditions, $\tau_{\mathrm{cem}}(0) = \mathrm{MVC}$. For this analysis, the MVC values for hip and knee were taken from Pereira (2009) as 245 Nm and 225 Nm, respectively. For the current configuration of the treadle pump, assuming 3 m of water depth and a treadling rate of 30 strokes per minute, i.e., 2 seconds per stroke, the endurance for sitting and standing positions is plotted in Fig. 4.20.

It is to be observed that even if one of the joints is unable to provide the required torque during either of the strokes, the user is considered fatigued. It can be seen that the endurance time of both joints is similar in the case of the standing position, whereas, in the sitting position, the endurance time of both joints varies significantly. The endurance times of sitting and standing positions in the current configuration of the RuTAG treadle pump for Fig. 4.20 are 1424 s and 1208 s, respectively. It is also to be noted that the results obtained are based on the assumption that the water source is at a depth of 3 m, and the user operates the pump continuously at a constant rate of 30 strokes per minute. Further, the dataset considered to estimate the mass-inertia properties of the lower limb corresponds to the 50th percentile male population (Pereira 2009). Therefore, in reality, the endurance times may vary based on the above factors.

4.2.3.5 Optimisation

The treadle pump of RuTAG IITD was designed such that it can be reconfigured and adjusted based on a user's requirements. The adjustable parts include the height and position of the seat and the position of the pedal. Furthermore, since the unit is

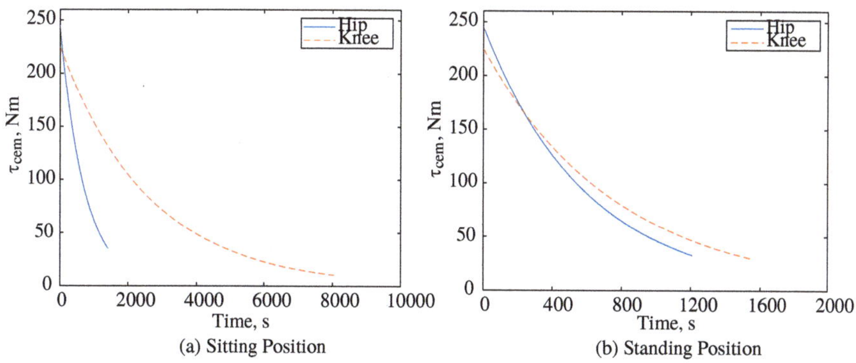

Fig. 4.20 Endurance for sitting and standing positions. (**a**) Sitting position (**b**) Standing position

not welded, individual parts, such as the handle height and position, and the lever lengths can be altered with minimum effort. In this section, one looks at how the treadle pump's configuration can be optimised in order to maximise the endurance time in the sitting position.

The three design variables considered for the optimisation problem are seat height h_s, the distance of the seat from pivot l_s, and the distance of the foot pedal from the pivot l_f. Ideally, the components such as the piston, cylinder, and plumbing parts can also be considered for optimisation. However, since standard parts are used in the fabrication of RuTAG's treadle pump, replacements for the optimised design may not be easily possible. Moreover, local manufacturing of these components is not practically feasible since they demand machining processes with higher accuracy, which is not expected to exist in a low-resource setting. On the other hand, the components, such as the seat and lever, do not require precision manufacturing and can be modified locally with ease. Hence, only the parameters like l_s, l_f, and h_s, as shown in Fig. 4.21, were considered for optimisation.

The objective function to be maximised is the endurance time t_{en}, which is obtained from the dynamic formulation explained in Sect. 4.2.3.4. The lower and upper bounds are obtained from the design constraints of the treadle pump. For instance, the longer the length of the treadle, the higher the amount of space required below the levers. Since this is a fixed system, increasing the space beyond a certain value is not feasible. In this case, the lever, at its lowest position, should not go below the base of the treadle pump. Similarly, while adjusting the seat height and position, it should be taken care that the foot pedal of the treadle is reachable to the user's leg. The algorithm used to optimise the variables is *Particle Swarm Optimisation*, like in Sect. 4.2.2.2. The results of the optimisation are shown in Table 4.6. The endurance time using the optimised configuration was found to be 2372 s as compared to 1424 s in the original configuration.

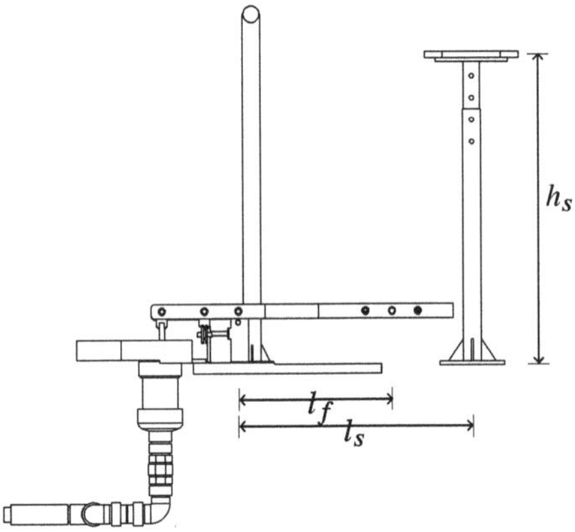

Fig. 4.21 Design variables for optimisation

Table 4.6 Design variables and bounds for the optimisation

Design variable (units)	Lower bound	Upper bound	Current value	Optimised value
h_s (m)	0.2	0.9	0.7	0.75
l_s (m)	0.2	0.8	0.656	0.45
l_f (m)	0.2	0.6	0.43	0.55

4.3 Summary

This chapter presented the design analysis of two BoP technologies: a wooden palanquin developed for the Yamunotri pilgrimage and a treadle pump. Both case studies demonstrate how engineering evaluation methods can be adapted to assess locally fabricated, resource-constrained technologies.

For the first case, structural analysis was carried out by modelling the palanquin considering the constraints such as local material and terrain conditions. It was found that the existing design was marginally safe, and a higher factor of safety was recommended.

The second case focused on the treadle pump. Forces acting on the treadles during both the power and return strokes were calculated, and structural analysis was carried out. It was found that the lever structure was safe for operation. For the dynamic analysis, aspects of clearance and ergonomics were considered. Clearance was considered in the revolute joint between the treadle and the connecting rod. The change in velocity and acceleration of the treadle, connecting rod, and piston was plotted. It was found that, due to multiple collisions, jerkiness was observed in the treadle and piston motions. The mass and inertia properties were optimised to smoothen the movement and reduce jerkiness. The ergonomics of the treadle pump operation was studied next. The joint torque values of the hip and knee were calculated using inverse dynamics. Muscle fatigue was used to estimate the duration for which the users can operate the pump continuously. Optimisation was used to prescribe the optimal configuration of the treadle pump for maximising the endurance time.

The study highlights the importance of integrating field data, user context, and constraints into the engineering analysis of BoP technologies. This approach ensured that the design recommendations arising from the analysis had greater relevance and were practically implementable. It was also seen that the effectiveness of a technology is measured not just by its performance but also by its appropriateness to the intended users.

References

Agarwal N, Oehler J, Brem A (2021) Constraint-based thinking: a structured approach for developing frugal innovations. IEEE Trans Eng Manag 68(3):739–751. https://doi.org/10.1109/TEM.2020.3042929

Ambrosio JA (2013) Multibody dynamics approaches to biomechanical applications to human motion tasks. In: Multibody system dynamics, robotics and control. Springer, Vienna, pp 259–289. https://doi.org/10.1007/978-3-7091-1289-2_16

Anderson FC, Pandy MG (2001) Dynamic optimization of human walking. J Biomech Eng 123(5):381–390

Ayoub M (1998) A 2-d simulation model for lifting activities. Comput Ind Eng 35(3-4):619–622

Bautista BN, Diaz Lankenau GF, Guitron SP, Jennings BD, Nechlani R, Tang AS, Tucker MR, Winter AG (2017) Design of an integrated cotton picking system for small-scale indian agriculture. In: Volume 4: 22nd design for manufacturing and the life cycle conference; 11th international conference on micro- and nanosystems. ASME, p V004T05A028. https://doi.org/10.1115/DETC2017-68132

Bazarevsky V, Grishchenko I, Raveendran K, Zhu T, Zhang F, Grundmann M (2020) BlazePose: on-device real-time body pose tracking. arXiv preprint arXiv:200610204

Bergmann JH, Noble A, Thompson M (2015) Why is designing for developing countries more challenging? Modelling the product design domain for medical devices. Proc Manuf 3(AHFE):5693–5698. https://doi.org/10.1016/j.promfg.2015.07.792

Bhat S, Saha SK, Gupta V (2019) Design analysis of the wooden palanquin used in Yamunotri Dham. In: 2019 IEEE global humanitarian technology conference (GHTC). IEEE, Piscataway, pp 1–5

Blajer W, Dziewiecki K, Mazur Z (2007) Multibody modeling of human body for the inverse dynamics analysis of sagittal plane movements. Multibody Syst Dyn 18(2):217–232. https://doi.org/10.1007/s11044-007-9090-2

Castellone R, Spada S, Sessa F, Cavatorta MP (2017) A simple multibody 2d-model for early postural checks in workplace design. Int J Appl Eng Res 12(23):13451–13461

Chakrabarti D (1997) Indian anthropometric dimensions for ergonomic design practice. National Institute of Design

Chakravarthy BK (2016) Light-weight Palkhi for heritage sites. http://www.idc.iitb.ac.in/chakku/Lightweight_Palkhi_for_Heritage_Sites.html

Chakravarthy BK (2019) Palkhi for Shri Mata Vaishno Devi shrine board, Katra, Jammu. http://www.idc.iitb.ac.in/chakku/Palkhi_Shri_Mata_Vaishno_Devi.html

Chaudhary H, Saha SK (2008) Dynamics and balancing of multibody systems, vol 37. Springer, Berlin

Copilusi C, Dumitru N, Margine A, Geonea I (2019) Dynamic analysis of a human ankle joint prosthesis. In: IOP conference series: materials science and engineering, vol 568. IOP Publishing, p 012077

De Groote F, Pipeleers G, Jonkers I, Demeulenaere B, Patten C, Swevers J, De Schutter J (2009) A physiology based inverse dynamic analysis of human gait: potential and perspectives. Comput Methods Biomech Biomed Eng 12(5):563–574

Ding J, Wexler AS, Binder-Macleod SA (2000) A predictive model of fatigue in human skeletal muscles. J Appl Physiol 89(4):1322–1332
Dul J, Weerdmeester B (2003) Ergonomics for beginners: a quick reference guide. CRC Press, Boca Raton
Fisher M (2006) Income is development: Kickstart's pumps help Kenyan farmers transition to a cash economy. Innov Technol Gov Glob 1(1):9–30
Green DW, Winandy JE, Kretschmann DE (1999) Mechanical properties of wood. wood handbook: wood as an engineering material. Forest Products Laboratory
Hertz H (1896) Miscellaneous papers. Macmillan, New York
Hill AV (1938) The heat of shortening and the dynamic constants of muscle. Proc Roy Soc Lond Series B-Biol Sci 126(843):136–195
Hill AV (1953) The mechanics of active muscle. Proc Roy Soc Lond Series B-Biol Sci 141(902):104–117
Jagtap S, Larsson A, Hiort V, Olander E, Warell A, Khadilkar P (2014) How design process for the base of the pyramid differs from that for the top of the pyramid. Design Stud 35(5):527–558. https://doi.org/10.1016/j.destud.2014.02.007
Kennedy J, Eberhart R (1995) Particle swarm optimization. In: Proceedings of ICNN'95—international conference on neural networks, vol 4, pp 1942–1948. https://doi.org/10.1109/ICNN.1995.488968
Lecomte C, Blanco E, Trompette P, Cholez C (2013) Towards a better frugal design using persona. In: 1st interdisciplinary innovation conference cooperating for innovation: devices for collective exploration
Liu JZ, Brown RW, Yue GH (2002) A dynamical model of muscle activation, fatigue, and recovery. Biophys J 82(5):2344–2359
Ma L, Chablat D, Bennis F, Zhang W (2009) A new simple dynamic muscle fatigue model and its validation. Int J Ind Ergon 39(1):211–220. https://doi.org/10.1016/j.ergon.2008.04.004
Ma R, Chablat D, Bennis F, Ma L (2012) Human muscle fatigue model in dynamic motions. In: Latest advances in robot kinematics, pp 349–356. https://doi.org/10.1007/978-94-007-4620-6_44
McComb C, Santaeufemia PS, Johnson NG, Shimada K (2014) Identifying technical and economic improvements to the MoneyMaker Hip pump through multi-objective optimization. In: IEEE global humanitarian technology conference (GHTC 2014). IEEE, Piscataway, pp 338–344. https://doi.org/10.1109/GHTC.2014.6970302
McComb C, Johnson NG, Gorman BT (2016) Scenario-based robustness analysis of optimized I.D.E.-style treadle pump designs. In: Volume 4: 21st design for manufacturing and the life cycle conference; 10th international conference on micro- and nanosystems. American Society of Mechanical Engineers, p V004T05A026. https://doi.org/10.1115/DETC2016-60127 https://asmedigitalcollection.asme.org/IDETC-CIE/proceedings/IDETC-CIE2016/50145/Charlotte. North Carolina, USA/258541
Nadadur G, Parkinson MB (2013) The role of anthropometry in designing for sustainability. Ergonomics 56(3):422–439. https://doi.org/10.1080/00140139.2012.718801
Ning H, Tan T, Wang L, Hu W (2004) Kinematics-based tracking of human walking in monocular video sequences. Image Vision Comput 22(5):429–441
Olorunnisola AO (2018) Design of structural elements with tropical hardwoods. Springer, Cham. https://doi.org/10.1007/978-3-319-65343-3
Pereira AF (2009) Development of a hill-type muscle model with fatigue for the calculation of the redundant muscle forces using multibody dynamics. Master's thesis, Instituto Superior Técnico, University of Lisbon
Pereira C, Malça J, Gaspar MC, Ventura F (2006) Human motion analysis in treadle pump devices. IFIP Int Feder Inform Process 221:135–145. https://doi.org/10.1007/978-0-387-36792-7_8
Portal R, Dias J, Soares CM (2006) Multibody models for vehicle accident reconstruction. In: ECCM 2006-III European conference on computational mechanics-solids, structures and coupled problems in engineering

Reinbolt JA, Schutte JF, Fregly BJ, Koh BI, Haftka RT, George AD, Mitchell KH (2005) Determination of patient-specific multi-joint kinematic models through two-level optimization. J Biomech 38(3):621–626

Rohmert W (1960) Determination of the recovery pause for static work of man. Internationale Zeitschrift Fur Angewandte Physiologie, Einschliesslich Arbeitsphysiologie 18:123–164

Saha SK (1999) Dynamics of serial multibody systems using the decoupled natural orthogonal complement matrices. J Appl Mech 66(4):986–996. https://doi.org/10.1115/1.2791809

Saha SK (2014) Introduction to robotics. Tata McGraw-Hill Education

Sanders MS, McCormick EJ (1987) Human factors in engineering and design. McGraw-Hill, New York

Sethi PK (1989) Technological choices in prosthetics and orthotics for developing countries. Prosthet Orthotics Int 13(3):117–124. https://doi.org/10.3109/03093648909079418

Silva MP, Ambrósio JA (2003) Solution of redundant muscle forces in human locomotion with multibody dynamics and optimization tools. Mech Based Design Struct Mach 31(3):381–411. https://doi.org/10.1081/SME-120022856

Silva MT, Ambrósio J, Pereira MS (1997) Biomechanical model with joint resistance for impact simulation. Multibody Syst Dyn 1(3):65–84. https://doi.org/10.1023/A:1009700405340

van der Bijl-Brouwer M, Dorst K (2017) Advancing the strategic impact of human-centred design. Design Stud 53:1–23. https://doi.org/10.1016/j.destud.2017.06.003

Varedi S, Daniali H, Dardel M, Fathi A (2015) Optimal dynamic design of a planar slider-crank mechanism with a joint clearance. Mech Mach Theory 86:191–200

Xia T, Law LAF (2008) A theoretical approach for modeling peripheral muscle fatigue and recovery. J Biomech 41(14):3046–3052

Zajac FE (1989) Muscle and tendon: properties, models, scaling, and application to biomechanics and motor control. Crit Rev Biomed Eng 17(4):359–411

Chapter 5
Recommendations

This chapter summarises the key findings reported in this book. The appropriate technology selection framework described in the book through case studies is summarised. Further, discussions about the structural and dynamic analyses are presented. Based on the observation of the authors, recommendations are made to effectively implement technological solutions in BoP settings.

5.1 Summary and Findings

This book provides a framework to select and analyse appropriate technologies for the resource-constrained environment, specifically for the users at the Base of the Economic Pyramid (BoP). It follows a human-centric approach to address the problem of technology adoption. One of the key reasons for the lack of adoption of technologies was found to be the improper selection and implementation of technologies.

Literature and field survey revealed that multiple variants of a single technology were often available for the users to purchase. Moreover, it was found that a particular variant that is successful in one context might not be successful in another context. This laid the requirement for a structured selection framework to choose the best possible alternative based on the user's context and requirements. To address this, a multi-attribute decision-making methodology, namely Technique for Order Preference by Similarity to Ideal Solution (TOPSIS), was implemented in Chap. 3.

A scenario-based selection procedure, where attribute weights can be extracted based on the user scenario and requirements, was demonstrated for the treadle pump, *Tulsi mala* bead-making machine, and vending cart. Further, the use of group decision-making and fuzzy attributes was also illustrated. From field data and the experience of the authors, attributes were identified in four categories:

S. Bhat et al., *Technologies for the Base of the Economic Pyramid*, Design Science and Innovation, https://doi.org/10.1007/978-981-95-2721-2_5

- **Financial:** Financial attributes were found to be a primary consideration for BoP users, given their limited financial resources. Unlike conventional markets, affordability in the BoP segment extends beyond the initial purchase cost. It includes factors such as payment models (e.g., Equated Monthly Instalments and rental services), maintenance costs, replacement costs for consumables, installation expenses, and after-sales support.
- **Contextual:** Contextual attributes help evaluate the feasibility of a technology within a given environment. Unlike standardised solutions, technologies for BoP markets must be context-specific, as user requirements can vary significantly depending on factors such as local infrastructure, climate conditions, user practices, and availability of supporting resources.
- **Performance:** Performance attributes serve as standard evaluation criteria for technology selection. These attributes assess parameters such as efficiency, reliability, durability, and output.
- **Ergonomic:** Ergonomic attributes are another critical factor influencing technology adoption in BoP settings. Poor ergonomic design can lead to user discomfort, decreased efficiency, and even rejection of the technology.

Given that different users have varying needs, scenarios were created based on the use cases of the technology. For instance, in the case of the treadle pump, scenarios were created based on the *type of terrain*, *type of irrigation*, and *type of locality*. Each scenario corresponds to a relative importance matrix, which was then used to extract the weights of attributes. A technology selection toolbox using C# was developed, using which any user can create scenarios and rank the technology alternatives. Although the main objective of the work carried out was to design a technology selection framework, key insights emerged from the case studies. For instance, the results of treadle pump ranking provide insights into how seemingly minor factors such as type of land and proximity to urban areas play an important role in the selection process. Similarly, in the case of the Tulsi mala device, important distinctions between men and women artisans were observed. One could observe that women gave a higher preference for comfort, whereas men preferred devices with higher operating speeds. It is also noteworthy that the rankings generated using the developed framework were also reflected in the user preferences in the field.

Two case studies were considered in Chap. 4 to demonstrate the aspects of structural and dynamic analyses that are important for a selected alternative chosen using the methodology prescribed in Chap. 2. Static structural analysis was carried out on a wooden palanquin used in Yamunotri *dham*. This case demonstrated how local factors, such as the availability of the type of wood and the geographical conditions of the field, influenced design decisions. Pinewood, being abundantly available locally, was being used to construct the palanquin. Moreover, the dimensions had a constraint to ensure that two palanquins, along with their porters, could easily pass through the terrain filled with sharp bends and turns. The results of the structural analyses showed that the palanquin was slightly underdesigned, and this was also validated by physical inspection of the used palanquin. One could observe the cracks and failures in the weaker areas of the palanquin. These insights emphasise the importance of structural optimisation in traditional designs which have not undergone formal design procedures.

To demonstrate the aspects of clearance and ergonomics, analyses were carried out on the treadle pump. Literature has highlighted the issue of maintaining the required manufacturing tolerances while fabricating in low-resource settings. In order to study the same, radial clearance was considered in one of the revolute joints of the treadle pump, and its effect on the pump performance was studied. Hertz's contact theory was used to model the elastic collisions between the journal and the bearing. The dynamics was modelled based on the DeNOC formulation. In order to minimise the jerkiness observed in the piston motion, mass-inertia properties of the links of the treadle pump were optimised. The objective was to minimise the distance between the journal and the bearing throughout the pumping stroke. It was seen that there was a considerable reduction in the jerkiness with the optimised mass-inertia properties of the links. Through the analysis it was demonstrated how the effects of joint clearance can be addressed using low-cost techniques such as mass redistribution.

The next aspect considered in Chap. 4 was ergonomics. It was seen in the literature that improper ergonomics is detrimental to the adoption of any technology. In the case of BoP, the effects are multiplied due to additional constraints such as purchasing power and livelihood dependency. In this book, the ergonomics of a treadle pump was studied by modelling and simulating the human-machine interaction. The lower limb of the human body was modelled as a multi-body system. The model was then interfaced with the multi-body model of the treadle pump using equivalent joint forces. Inverse dynamics was performed with treadle motion being the input. Joint torques were obtained at the hip and knee joints of the human body. The concept of muscle fatigue was used to estimate the maximum time duration for which a user can operate the treadle pump continuously. The estimation was done for both the sitting and standing position. For the standing position, a Python-based AI tool called *mediapipe* was used to estimate the joint angles. Using optimisation, the optimal configuration of the treadle pump was suggested to maximise the endurance time. This study demonstrated how ergonomic refinements can enhance user experience, ultimately leading to greater adoption of the technology. The findings also highlight the importance of customising the designs as per the working condition and capabilities of the users, reinforcing the need for human-centric technology development.

Overall, this book addresses the issue of technological sustainability. Through the topics such as the selection of appropriate technology and optimisation for better ergonomics and user experience, the book emphasises the usage of modern tools for technology selection and detailed design analyses, which can lead to better and proper utilisation of the technologies developed for the base of the pyramid. Although environmental sustainability is out of the scope of the current research, the outcome of the current research leads to the socio-economic sustainability of the rural areas by providing appropriate decision-making tools and design recommendations for resource-constrained environments.

Key recommendations based on the findings of the research are presented in terms of Technology Selection, Design and Analysis, and Technology Implementation.

5.1.1 Technology Selection

- **Adopt a structured decision-making framework:** Technology developers and policymakers should employ structured frameworks such as the TOPSIS-based selection methodology to ensure that the chosen technology aligns with the needs of Base of the Pyramid (BoP) users. Moreover, multiple stakeholders should be involved in the decision-making process so that diverse perspectives are considered.
- **Implement scenario-based selection process:** User context is paramount when it comes to the BoP segment. Practitioners should appropriately identify technology attributes and assign weights based on the usage scenario and context before carrying out technology selection. The scenario-based approach demonstrated for the treadle pump, Tulsi mala bead-making machine, and vending cart should be expanded to other BoP technologies.

5.1.2 Design and Analysis

- **Incorporate optimisation and ergonomic considerations:** Traditional technologies should undergo formal structural analysis to identify areas of improvement. For example, the structural deficiencies found in the Yamunotri palanquin highlight the need for optimisation in load-bearing structures. Furthermore, ergonomics should be a major aspect of design and optimisation. The findings from the treadle pump study suggest that addressing ergonomic deficiencies can significantly improve performance and usability.
- **Implement low-cost solutions to address performance issues:** Whenever possible, cost-effective solutions should be implemented to improve the performance of the technology. For instance, mass redistribution was preferred over using high-precision manufacturing to mitigate the effects of joint clearances in treadle pumps.

5.1.3 Technology Implementation

- **Engage users in the design process:** Participatory design approaches should be encouraged. Involvement of all the stakeholders during each stage, be it technology selection, design and analysis, or implementation, will ensure sustainable technology adoption.

- **Provide after-sales support and training:** Ensuring that the designed technology has good support in terms of service, spare parts, and maintenance will instil confidence in the users to adopt the technology. Training programmes to upskill users and local manufacturers also help in technology adoption, ensure proper usage, and facilitate wider dissemination.

5.2 Limitations and Scope of Future Work

In this book, scenario-based selection methodology was introduced. The methodology was used to rank the technology alternatives. The cases demonstrated in this thesis were the technologies that were already developed and available in the market. The scenario-based selection can also be used in initial design stages, such as concept evaluations, so that the developed solutions have a greater chance of creating the intended impact. Moreover, the impact of the selection methodology can be studied only after implementing the same in the field. In the case of *Tulsi mala* bead-making device, although field validation could be done in some cases, thorough field testing is required to validate the rankings. Similarly, for the street vending carts, the current study only demonstrated how the ranking methodology could be applied to qualitative attributes. Using user data and field-based decision-makers can give more realistic scenario-based rankings.

The dynamic analysis of the treadle pump mechanism showed the effects of large clearances on the pump performance in terms of jerkiness observed in the piston movement. In order to reduce the negative impact of such clearances, optimal mass-inertia properties were suggested based on the results of optimisation. It is to be noted that the realisation of the prescribed mass-inertia properties requires additional techniques, such as shape optimisation, which can be taken up in the future. Note that the optimal mass-inertia properties were a result of the optimisation problem with limited constraints. Additional constraints, such as operating force, factor of safety, and component stress values, can also be included in future for the optimisation.

The ergonomic study undertaken in this book considered a two-dimensional model of the treadle pump and the human body. Only the lower limb was considered for the study. A three-dimensional model of both the treadle pump and the human body might give additional insights into the human-machine interaction of the treadle pump. The muscle fatigue model considered in the work was for ideal conditions. In real life, particularly in the case of low-resource settings, factors such as heat, nutrition, etc. would affect the endurance of the user. Hence, the inclusion of those with accurate physiological data of the target community can help the designers to develop better solutions.

Looking at the broader perspective, one can utilise the findings of this research to move towards a unified framework that addresses overall product development for the BoP. Although the technical aspects of selection and analysis are demonstrated in this book, the actual implications can only be studied once it is implemented in the field for a considerable time. Moreover, from a holistic view, one has to consider constraints such as low cost-low volume requirements and lower purchasing power of the target population. Combining this with innovative business models could boost rural innovation and create entrepreneurship opportunities.

Appendix A
Details of Treadle Pumps

Brief details of various treadle pumps are given below.

A.1 Ajay Industrial Treadle Pump

Ajay Industrial Corporation Ltd. is a manufacturer of hand pumps and treadle pumps based in Ghaziabad, India. They export pumps to African nations, including Zambia and Ethiopia.

The treadle pump manufactured by them is made of galvanised steel and makes use of a cam-chain mechanism for reciprocating motion. This is a pressure pump, and the water can be delivered at a pressure so as to reach higher levels. More details and specifications can be found on their website https://www.ajayindustrial.com/.

A.2 RuTAG-IITD Treadle Pump

Rural Technology Action Group (RUTAG) was conceived by the Office of the Principal Scientific Adviser to the Government of India to provide a higher level of Science and Technology inputs to problems pertaining to rural technologies. RuTAG at IIT Delhi is one such centre and was established in 2009.

The RuTAG-IITD treadle pump is made of mild steel. It consists of a rocker-lever mechanism to facilitate the reciprocating motion of two levers. This is a suction-type treadle pump which means that the pumped water is delivered at the atmospheric pressure through a delivery channel. More details can be found on the website http://rutag.iitd.ac.in.

S. Bhat et al., *Technologies for the Base of the Economic Pyramid*, Design Science and Innovation, https://doi.org/10.1007/978-981-95-2721-2

A.3 Ecoflo Treadle Pump

Ecoflo India produces micro-irrigation kits and pumping systems. It was established in 1991 and is based out of Maharashtra, India. Their products are sold in over 45 countries worldwide.

The cylinders and manifolds of Ecoflo treadle pumps are made of injection-moulded thermoplastic. The frame is made of powder-coated steel. The pump consists of a rope-pulley mechanism for reciprocation motion. More information can be found at https://ecofloindia.com/.

A.4 RuKart Treadle Pump

RuKart is an IIT Bombay-based agri-tech start-up with an aim to develop affordable technologies for small and marginal farmers. Their current lineup of products includes manual pumps, cold storage, crop protection devices, and water management technologies.

RuKart manufactures metallic treadle pumps. The pumps are capable of delivering water at a pressure and can be used to irrigate medium-sized farmlands. More details can be found at http://rukart.org.

A.5 Moneymaker Max

Moneymaker Max is a treadle pump developed by Kickstart International, a non-profit social enterprise. They mostly cater to farmers across Africa. Moneymaker Max is a metallic treadle pump, which is a pressure-type pump. Details can be found on the website https://kickstart.org/.

Appendix B
Equations of Motion

Referring to Fig. 4.8, the equations of motion can be derived using the DeNOC formulation as follows.

B.1 Generalised Inertia Matrix

For the subsystem I, the elements of the 2×2 generalised inertia matrix (GIM) can be obtained as

$$i_{22} = \mathbf{p}_2^T \widetilde{\mathbf{M}}_2 \mathbf{B}_{22}\, \mathbf{p}_2 \tag{B.1}$$

$$i_{21} = i_{12} = \mathbf{p}_2^T \widetilde{\mathbf{M}}_2 \mathbf{B}_{21}\, \mathbf{p}_1 \tag{B.2}$$

$$i_{11} = \mathbf{p}_1^T \widetilde{\mathbf{M}}_1 \mathbf{p}_1 \tag{B.3}$$

where

$$\mathbf{p}_i = \begin{bmatrix} \mathbf{e}_i \\ \mathbf{e}_i \times \mathbf{d}_i \end{bmatrix}; \mathbf{B}_{ij} = \begin{bmatrix} \mathbf{1} & \mathbf{O} \\ -(\mathbf{r}_j + \mathbf{d}_i) \times \mathbf{1} & \mathbf{1} \end{bmatrix}; \text{ and } \widetilde{\mathbf{M}}_i = \mathbf{M}_i + \mathbf{B}_{i+1,i}^T \widetilde{\mathbf{M}}_{i+1} \mathbf{B}_{i+1} \tag{B.4}$$

Since both the joints in subsystem I are planar revolute, $\mathbf{e}_i = [0, 0, 1]^T$. For an n link system,

$$\widetilde{\mathbf{M}}_n = \mathbf{M}_n = \begin{bmatrix} \mathbf{I}_n & \mathbf{O} \\ \mathbf{O} & m_n \mathbf{1} \end{bmatrix} \tag{B.5}$$

The inertia tensors $\mathbf{I}_i$ $(i = 1, 2)$ obtained using the geometric properties of the link are

S. Bhat et al., *Technologies for the Base of the Economic Pyramid*, Design Science and Innovation, https://doi.org/10.1007/978-981-95-2721-2

$$\mathbf{I}_1 = \frac{3m_1 l_1^2}{32}\begin{bmatrix}0&0&0\\0&1&0\\0&0&1\end{bmatrix}; \mathbf{I}_2 = \frac{7m_2 l_2^2}{90}\begin{bmatrix}0&0&0\\0&1&0\\0&0&1\end{bmatrix} \tag{B.6}$$

where m_i and l_i are the mass and the length of links. Using Eqs. B.1–B.6, the elements of the GIM can be obtained as

$$i_{22} = \frac{7m_2 a_2^2}{90} + m_2 d_2^2 \tag{B.7}$$

$$i_{21} = i_{12} = \frac{7m_2 a_2^2}{90} + m_2 d_2^2 + m_2 d_2 a_1 \cos\theta_2 \tag{B.8}$$

$$\begin{aligned} i_{11} = & \frac{7m_2 a_2^2}{90} + (m_1 + m_2) d_1^2 + m_2 d_2^2 + \frac{3m_1 l_1^2}{32} \\ & + m_2 r_1^2 + 2 d_1 m_2 r_1 + 2 m_2 d_2 a_1 \cos\theta_2 \end{aligned} \tag{B.9}$$

The GIM for subsystem I can then be written as

$$\mathbf{I}_{\mathrm{I}} = \begin{bmatrix} i_{11} & i_{12} \\ i_{21} & i_{22} \end{bmatrix} \tag{B.10}$$

For subsystem II, GIM is written as $\mathbf{I}_{\mathrm{II}} = m_2$.

B.2 Matrix for Convective Inertia

The elements of the matrix for convective inertia (MCI) can be obtained as

$$c_{22} = \mathbf{p}_2^T[(\mathbf{B}_{32}^T\widetilde{\mathbf{H}}_{32} + \mathbf{B}_{22}^T(\widetilde{\mathbf{M}}_2\boldsymbol{\Omega}_2 + \widetilde{\mathbf{M}}_2')]\mathbf{p}_2 \tag{B.11}$$

$$c_{21} = \mathbf{p}_2^T[\widetilde{\mathbf{H}}_{21} + \widetilde{\mathbf{M}}_2\mathbf{B}_{21}\boldsymbol{\Omega}_1 + \widetilde{\mathbf{M}}_2'\mathbf{B}_{21}]\mathbf{p}_1 \tag{B.12}$$

$$c_{12} = \mathbf{p}_1^T[(\mathbf{B}_{31}^T\widetilde{\mathbf{H}}_{32} + \mathbf{B}_{21}^T(\widetilde{\mathbf{M}}_2\boldsymbol{\Omega}_2 + \widetilde{\mathbf{M}}_2')]\mathbf{p}_2 \tag{B.13}$$

$$c_{11} = \mathbf{p}_1^T[(\mathbf{B}_{21}^T\widetilde{\mathbf{H}}_{21} + \mathbf{B}_{11}^T(\widetilde{\mathbf{M}}_1\boldsymbol{\Omega}_1 + \widetilde{\mathbf{M}}_1')]\mathbf{p}_1 \tag{B.14}$$

Here, $\widetilde{\mathbf{H}}_{ij} = \widetilde{\mathbf{M}}_i\dot{\mathbf{B}}_{\mathbf{ij}} + \mathbf{B}_{i+1,i}^T\widetilde{\mathbf{H}}_{i+1,i}$, and for an n link system, $\widetilde{\mathbf{H}}_{n+1,n} = \mathbf{O}$; $\widetilde{\mathbf{M}}_i' = \mathbf{M}_i' + \mathbf{B}_{i+1,i}^T\widetilde{\mathbf{M}}_{i+1}'\mathbf{B}_{i+1,i}$, where $\widetilde{\mathbf{M}}_i' = \mathbf{W}_i\mathbf{M}_i$. $\mathbf{W}_i$ and $\boldsymbol{\Omega}_i$ are given by

$$\mathbf{W}_i = \begin{bmatrix}\dot{\theta}_i\mathbf{e}\times\mathbf{1} & \mathbf{O}\\ \mathbf{O} & \mathbf{O}\end{bmatrix}; \ \boldsymbol{\Omega}_i = \begin{bmatrix}\dot{\theta}_i\mathbf{e}\times\mathbf{1} & \mathbf{O}\\ \mathbf{O} & \dot{\theta}_i\mathbf{e}\times\mathbf{1}\end{bmatrix} \tag{B.15}$$

Using Eqs. B.11–B.15, the MCI elements can be written as

$$c_{22} = 0 \tag{B.16}$$

$$c_{21} = a_1 d_2 m_2 \sin\theta_2 \dot{\theta}_1 \tag{B.17}$$

$$c_{12} = -a_1 d_2 m_2 \sin\theta_2 (\dot{\theta}_1 + \dot{\theta}_2) \tag{B.18}$$

$$c_{11} = -a_1 d_2 m_2 \sin\theta_2 \dot{\theta}_2 \tag{B.19}$$

The MCI for subsystem I can be written as

$$\mathbf{C}_{\mathrm{I}} = \begin{bmatrix} c_{11} & c_{12} \\ c_{21} & c_{22} \end{bmatrix} \tag{B.20}$$

The vector of convective inertia (VCI) terms $\mathbf{h}$ is then obtained as

$$\mathbf{h}_{\mathrm{I}} = \mathbf{C}_{\mathrm{I}} \dot{\boldsymbol{\theta}}_{\mathrm{I}} \tag{B.21}$$

where $\boldsymbol{\theta}_{\mathrm{I}} = [\theta_1, \theta_2]^T$

B.3 Constraint Jacobians

The constraint equations can be written as follows:

$$a_1 \cos\theta_1 + a_2 \cos(\theta_1 + \theta_2) = 0.225 \tag{B.22}$$

$$a_1 \sin\theta_1 + a_2 \sin(\theta_1 + \theta_2) = b \tag{B.23}$$

Differentiating the constraint equations and writing them in the form of $\mathbf{J}\dot{\boldsymbol{\theta}}$, we can obtain the Jacobian as

$$\mathbf{J} = \begin{bmatrix} -(a_1 \sin\theta_1 + a_2 \sin(\theta_1 + \theta_2)) & -a_2 \sin(\theta_1 + \theta_2) & 0 \\ a_1 \cos\theta_1 + a_2 \cos(\theta_1 + \theta_2) & a_2 \cos(\theta_1 + \theta_2) & -1 \end{bmatrix} \tag{B.24}$$

B.4 Generalised Force Vectors

The vector of external forces $\boldsymbol{\tau}$ can be written for both the subsystems as

$$\boldsymbol{\tau}_{\mathrm{I}} = \begin{bmatrix} \mathbf{p}_1^T \widetilde{\mathbf{w}}_1^E \\ \mathbf{p}_2^T \widetilde{\mathbf{w}}_2^E \end{bmatrix}; \quad \boldsymbol{\tau}_{\mathrm{II}} = \mathbf{p}_3^T \widetilde{\mathbf{w}}_3^E \tag{B.25}$$

where $\widetilde{\mathbf{w}}_i^E = \mathbf{w}_i^E + \mathbf{B}_{i+1,i}^T \widetilde{\mathbf{w}}_{i+1}^E$, and for an n-link system, $\widetilde{\mathbf{w}}_n^E = \mathbf{w}_n^E$, where $\mathbf{w}^E$ contains the external forces for each link.

The combined equations of motion for both subsystems can be written as

$$\begin{bmatrix} \mathbf{I}_{\mathrm{I}} & \mathbf{O} & -\mathbf{J}_{\mathrm{I}}^{T} \\ \mathbf{O} & \mathbf{I}_{\mathrm{II}} & -\mathbf{J}_{\mathrm{II}}^{T} \\ -\mathbf{J}_{\mathrm{I}} & -\mathbf{J}_{\mathrm{II}} & \mathbf{O} \end{bmatrix} \begin{bmatrix} \ddot{\boldsymbol{\theta}}_{\mathrm{I}} \\ \ddot{\boldsymbol{\theta}}_{\mathrm{II}} \\ \boldsymbol{\lambda} \end{bmatrix} = \begin{bmatrix} \boldsymbol{\tau}_{\mathrm{I}} - \mathbf{h}_{\mathrm{I}} \\ \boldsymbol{\tau}_{\mathrm{II}} - \mathbf{h}_{\mathrm{II}} \\ \dot{\mathbf{J}}_{\mathrm{I}}\dot{\boldsymbol{\theta}}_{\mathrm{I}} + \dot{\mathbf{J}}_{\mathrm{II}}\dot{\boldsymbol{\theta}}_{\mathrm{II}} \end{bmatrix} \tag{B.26}$$

Appendix C
Modelling Clearance

With reference to Fig. 4.10, subsystem approach is followed to model clearance in the treadle pump. Clearance is considered in the revolute joint R. As a result, the treadle forms a one-link subsystem I. The PR mechanism consisting of the piston and the connecting rod forms subsystem II.

C.1 Generalised Inertia Matrix

For subsystem I, the Generalised Inertia Matrix (GIM) can be obtained as

$$\mathbf{I}_{\mathrm{I}} = \mathbf{p}_1^T \mathbf{M}_1 \mathbf{p}_1 \tag{C.1}$$

where

$$\mathbf{p}_1 = \begin{bmatrix} \mathbf{e}_1 \\ \mathbf{e}_1 \times \mathbf{d}_1 \end{bmatrix}; \mathbf{e}_1 = [0, 0, 1]^T; \mathbf{M}_1 = \begin{bmatrix} \mathbf{I}_1 & \mathbf{O} \\ \mathbf{O} & m_1\mathbf{1} \end{bmatrix} \tag{C.2}$$

and

$$\mathbf{I}_1 = \frac{3m_1 l_1^2}{32} \begin{bmatrix} 0 & 0 & 0 \\ 0 & 1 & 0 \\ 0 & 0 & 1 \end{bmatrix} \tag{C.3}$$

Using Eqs. C.1–C.3, GIM can be obtained as

$$\mathbf{I}_{\mathrm{I}} = \frac{3m_1 l_1^2}{32} + m_1 d_1^2 \tag{C.4}$$

S. Bhat et al., *Technologies for the Base of the Economic Pyramid*, Design Science and Innovation, https://doi.org/10.1007/978-981-95-2721-2

For subsystem II, elements of the 2 × 2 GIM can be obtained as

$$i_{33} = \mathbf{p}_3^T \widetilde{\mathbf{M}}_3 \mathbf{B}_{33}\, \mathbf{p}_3 \tag{C.5}$$

$$i_{32} = i_{23} = \mathbf{p}_3^T \widetilde{\mathbf{M}}_3 \mathbf{B}_{32}\, \mathbf{p}_2 \tag{C.6}$$

$$i_{22} = \mathbf{p}_2^T \widetilde{\mathbf{M}}_2 \mathbf{p}_2 \tag{C.7}$$

For prismatic joints,

$$\mathbf{p}_2 = \begin{bmatrix} \mathbf{0} \\ \mathbf{e}_2 \end{bmatrix};\ \mathbf{e}_2 = [0, 1, 0]^T \tag{C.8}$$

The terms can be obtained as shown in Sect. B.1.

$$i_{33} = \frac{7 m_3 a_3^2}{90} + m_3 d_3^2 \tag{C.9}$$

$$i_{32} = i_{23} = m_3 d_3 \cos\theta_3 \tag{C.10}$$

$$i_{22} = m_2 + m_3 \tag{C.11}$$

The GIM for subsystem II can then be written as

$$\mathbf{I}_{\mathrm{II}} = \begin{bmatrix} i_{22}\ i_{23} \\ i_{32}\ i_{33} \end{bmatrix} \tag{C.12}$$

C.2 Matrix for Convective Inertia

The matrix for convective inertia (MCI) terms can be obtained as explained in Sect. B.2. The MCI terms for subsystem II can be obtained as

$$c_{22} = 0 \tag{C.13}$$

$$c_{21} = -m_3 d_3 \sin\theta_3 \dot{\theta}_3 \tag{C.14}$$

$$c_{12} = 0 \tag{C.15}$$

$$c_{11} = 0 \tag{C.16}$$

The vector of convective inertia (VCI) terms $\mathbf{h}$ is then obtained as

$$\mathbf{h}_{\mathrm{II}} = \mathbf{C}_{\mathrm{II}} \dot{\theta}_{\mathrm{II}} \tag{C.17}$$

where $\theta_{\mathrm{II}} = [b, \theta_3]^T$

C.3 Generalised Force Vector

The contact forces generated during impact are included in the generalised force vector. The normal force F_n can be calculated using Eqs. 4.4–4.6. Further, if δ_x and δ_y are the relative displacements of the journal with respect to the bearing, the components of the normal force can be obtained as

$$F_x = F_n \cos\phi \tag{C.18}$$

$$F_y = F_n \sin\phi \tag{C.19}$$

where

$$\phi = \tan^{-1}\frac{\delta_y}{\delta_x} \tag{C.20}$$

Since the forces are acting at the end of the link, they must be transferred to the mass centre before writing the force vectors. Using equivalent forces, the vector of external forces $\boldsymbol{\tau}$ can then be written for both the subsystems as

$$\boldsymbol{\tau}_{\mathrm{I}} = \mathbf{p}_1^T \widetilde{\mathbf{w}}_1^E; \;\; \boldsymbol{\tau}_{\mathrm{II}} = \begin{bmatrix} \mathbf{p}_2^T \widetilde{\mathbf{w}}_2^E \\ \mathbf{p}_3^T \widetilde{\mathbf{w}}_3^E \end{bmatrix} \tag{C.21}$$

The vector of gravity terms can be obtained as

$$\boldsymbol{\gamma}_{\mathrm{I}} = \mathbf{p}_1^T \mathbf{g}_1; \;\; \boldsymbol{\gamma}_{\mathrm{II}} = \begin{bmatrix} \mathbf{p}_2^T \mathbf{g}_2 \\ \mathbf{p}_3^T \mathbf{g}_3 \end{bmatrix} \tag{C.22}$$

where $\mathbf{g} = [0, g, 0]^T$. The combined equations of motion can be written as

$$\begin{bmatrix} \mathbf{I}_{\mathrm{I}} & \mathbf{O} \\ \mathbf{O} & \mathbf{I}_{\mathrm{II}} \end{bmatrix} \begin{bmatrix} \ddot{\boldsymbol{\theta}}_{\mathrm{I}} \\ \ddot{\boldsymbol{\theta}}_{\mathrm{II}} \end{bmatrix} = \begin{bmatrix} \boldsymbol{\tau}_{\mathrm{I}} + \boldsymbol{\gamma}_{\mathrm{I}} \\ \boldsymbol{\tau}_{\mathrm{II}} - \mathbf{h}_{\mathrm{II}} + \boldsymbol{\gamma}_{\mathrm{II}} \end{bmatrix} \tag{C.23}$$

Appendix D
Inverse Dynamics for Ergonomic Analysis

Joint reactions were obtained using inverse dynamics. For the sitting position, refer to Fig. 4.16b. For consistency, subsystem II is renumbered to links 1 and 2 instead of links 3 and 2. Thus, the joint reactions to be calculated are at O_1 (hip) and O_2 (knee).

D.1 Twist and Twist Rates

The twists of the link can be obtained as

$$\mathbf{t}_1 = \mathbf{p}_1\dot{\theta}_1 \tag{D.1}$$

$$\mathbf{t}_2 = \mathbf{A}_{21}\mathbf{t}_1 + \mathbf{p}_2\dot{\theta}_2 \tag{D.2}$$

where $\mathbf{p}_i$ is $[\mathbf{e}_i, \mathbf{0}]^T$ for revolute joint, and $\mathbf{A}_{21}$ is the twist propagation matrix given by

$$\mathbf{A}_{21} = \begin{bmatrix} \mathbf{1} & \mathbf{O} \\ \tilde{\mathbf{a}}_{21} & \mathbf{1} \end{bmatrix} \tag{D.3}$$

where $\tilde{\mathbf{a}}_{21}$ is the 3×3 skew-symmetric matrix, and $\mathbf{a}_{21} = -\mathbf{a}_{12}$, where $\mathbf{a}_{12}$ is the vector from the origin of link 1 to the origin of link 2. The twist rates $\dot{\mathbf{t}}_1$ and $\dot{\mathbf{t}}_2$ are obtained by differentiating Eqs. D.1 and D.2.

S. Bhat et al., *Technologies for the Base of the Economic Pyramid*, Design Science and Innovation, https://doi.org/10.1007/978-981-95-2721-2

D.2 Joint Wrenches

The joint wrenches are calculated as

$$\begin{bmatrix} \mathbf{w}_{01} \\ \mathbf{w}_{12} \end{bmatrix} = \begin{bmatrix} \mathbf{1} & \mathbf{A}'_{12} \\ \mathbf{O} & \mathbf{1} \end{bmatrix} \begin{bmatrix} \mathbf{w}'_1 \\ \mathbf{w}'_2 \end{bmatrix} \tag{D.4}$$

It is to be noted that the calculations are done inwards, i.e. $\mathbf{w}_{12}$ is calculated first.

$$\mathbf{w}_{12} = \mathbf{w}'_2 = \mathbf{w}_2^* - \mathbf{w}_2^e \tag{D.5}$$

where

$$\mathbf{w}_2^e = \begin{bmatrix} \widetilde{\mathbf{a}}_2 \mathbf{f}_2^e + \widetilde{\mathbf{d}}_2 \mathbf{f}_2^g \\ \mathbf{f}_2^e + \mathbf{f}_2^g \end{bmatrix} \tag{D.6}$$

Here $\mathbf{f}_2^e$ represents the external force vector acting at the end of link 2, and $\mathbf{f}_2^e$ represents the gravity terms. $\mathbf{d}_2$ is the vector from the origin of link 2 to the mass centre of link 2. The inertia wrench of link 2 $\mathbf{w}_2^*$ is given by

$$\mathbf{w}_2^* = \mathbf{M}_2 \dot{\mathbf{t}}_2 + \mathbf{W}_2 \mathbf{M}_2 \mathbf{E}_2 \mathbf{t}_2 \tag{D.7}$$

The 6×6 mass matrix $\mathbf{M}_2$ is obtained as

$$\mathbf{M}_2 = \begin{bmatrix} \mathbf{I}_2 & m_2 \widetilde{\mathbf{d}}_2 \\ -m_2 \widetilde{\mathbf{d}}_2 & m_2 \mathbf{1} \end{bmatrix} \tag{D.8}$$

where m_2 is the mass of link 2, and $\mathbf{I}_2$ is the inertia tensor of link 2 about its origin, which can be obtained as $\mathbf{I}_2 = \mathbf{I}_2^c - m_2 \widetilde{\mathbf{d}}_2^2$. The angular velocity matrix $\mathbf{W}_2$ is given by

$$\mathbf{W}_2 = \begin{bmatrix} \widetilde{\boldsymbol{\omega}}_2 & \mathbf{O} \\ \mathbf{O} & \widetilde{\boldsymbol{\omega}}_2 \end{bmatrix}; \boldsymbol{\omega}_2 = [0\ 0\ \dot{\theta}_{12}]^T \tag{D.9}$$

and

$$\mathbf{E}_2 = \begin{bmatrix} \mathbf{1} & \mathbf{O} \\ \mathbf{O} & \mathbf{O} \end{bmatrix} \tag{D.10}$$

The constraint wrench at joint O_2 (knee), $\mathbf{w}_{12}$, is obtained in the form of

$$\mathbf{w}_{12} = \begin{bmatrix} \tau_{12x} \\ \tau_{12y} \\ \tau_{12z} \\ f_{12x} \\ f_{12y} \\ f_{12z} \end{bmatrix} \tag{D.11}$$

from which the joint torques and reactions can be extracted. Similarly, constraint wrench at O_1 is obtained as

$$\mathbf{w}_{01} = \mathbf{w}_1^{'} + \mathbf{A}_{12}^{'}\mathbf{w}_2^{'} \tag{D.12}$$

The *wrench propagation matrix* $\mathbf{A}_{i,i+1}^{'}$ is given by

$$\mathbf{A}_{i,i+1}^{'} = \begin{bmatrix} \mathbf{1} & \widetilde{\mathbf{a}}_{i,i+1} \\ \mathbf{O} & \mathbf{1} \end{bmatrix} \tag{D.13}$$

where $\mathbf{a}_{i,i+1}$ is the vector from the origin of ith link to the origin of $(i+1)$th link. Further,

$$\mathbf{w}_1^{'} = \mathbf{w}_1^{*} - \mathbf{w}_1^{e}\ ;\ \ \mathbf{w}_1^{*} = \mathbf{M}_1\dot{\mathbf{t}}_1 + \mathbf{W}_1\mathbf{M}_1\mathbf{E}_1\mathbf{t}_2\ ;\ \ \mathbf{w}_1^{e} = \begin{bmatrix} \widetilde{\mathbf{d}}_1\mathbf{f}_1^{g} \\ \mathbf{f}_1^{g} \end{bmatrix} \tag{D.14}$$

Using similar approaches as shown in Eqs. D.8–D.10, the constraint wrench at joint O_1 (hip) is obtained in the form of

$$\mathbf{w}_{01} = \begin{bmatrix} \tau_{01x} \\ \tau_{01y} \\ \tau_{01z} \\ f_{01x} \\ f_{01y} \\ f_{01z} \end{bmatrix} \tag{D.15}$$

Thus, reaction forces and torques at both the joints O_1 (hip) and O_2 (knee) can be calculated.

MIX
Papier aus verantwortungsvollen Quellen
Paper from responsible sources
FSC® C105338

If you have any concerns about our products,
you can contact us on
ProductSafety@springernature.com

In case Publisher is established outside the EU,
the EU authorized representative is:
Springer Nature Customer Service Center GmbH
Europaplatz 3, 69115 Heidelberg, Germany

Printed by Libri Plureos GmbH
in Hamburg, Germany